Resource Alloc

for the New Defense Strategy

The DynaRank Decision-Support System

Richard J. Hillestad
Paul K. Davis

Prepared for the
Office of the Secretary of Defense

RAND *National Defense Research Institute*

The research described in this report was sponsored by the Office of the Secretary of Defense (OSD). The research was conducted in RAND's National Defense Research Institute, a federally funded research and development center supported by the OSD, the Joint Staff, the unified commands, and the defense agencies under Contract DASW01-95-C-0059.

Library of Congress Cataloging-in-Publication Data

Hillestad, R. J. (Richard John), 1942– .
Resource allocation for the new defense strategy : the DynaRank decision-support system / Richard J. Hillestad and Paul K. Davis.
p. cm.
"Prepared for the Office of the Secretary of Defense by RAND's National Defense Research Institute."
"MR-996-OSD."
Includes bibliographical references (p.).
ISBN 0-8330-2652-6
1. Strategy. 2. Military planning—United States—Decision making—Data processing. 3. United States. Dept. of Defense—Decision making—Data processing. I. Davis, Paul K., 1943– . II. United States. Dept. of Defense. Office of the Secretary of Defense. III. National Defense Research Institute (U. S.). IV. Title.
U162.H55 1998
355 ' .033073—dc21 98-27925
CIP

Published 1998 by RAND
1700 Main Street, P.O. Box 2138, Santa Monica, CA 90407-2138
1333 H St., N.W., Washington, D.C. 20005-4707
RAND URL: http://www.rand.org/
To order RAND documents or to obtain additional information, contact Distribution Services: Telephone: (310) 451-7002; Fax: (310) 451-6915; Internet: order@rand.org

PREFACE

This report describes a decision-support system called DynaRank. It illustrates DynaRank's use in defense planning with examples based on work for the Quadrennial Defense Review and follow-up research in 1997. It is our hope that a version of this methodology, which explicitly links program-level choices to higher-level strategy, will become part of the Department of Defense Planning, Programming and Budgetary System (PPBS). We believe that a methodology that considers all of the components of the new defense strategy—respond, shape, and prepare now—will help provide rationale for making difficult programmatic choices if necessary.

This report and DynaRank should not only be of interest to military analysts but also to policy analysts and other researchers in general, because of DynaRank's broader applicability to evaluating policy options.

This document describes the DynaRank tool in a fair amount of detail, including an appendix that can be used as a tutorial. DynaRank is a Microsoft® Excel workbook available for the Macintosh and an IBM-compatible computer.

The research reported here was conducted as part of RAND's "Planning Future Forces" project, a cross-cutting effort sponsored by the advisory board of RAND's National Defense Research Institute, a federally funded research and development center supported by the Office of the Secretary of Defense, the Joint Staff, the unified commands, and the defense agencies.

Questions about DynaRank or its application should be directed to the authors: Richard_Hillestad@rand.org and Paul_Davis@rand.org.

CONTENTS

FIGURES

TABLES

SUMMARY

The purpose of this report is to describe a decision-support system called DynaRank, which is designed to assist the Department of Defense's (DoD's) development and updating of the U.S. defense program in a way that is explicitly consistent with the multiple objectives of the new strategy: shape, respond, and prepare now. We have been motivated by the observation that the current DoD Planning, Programming and Budgetary System (PPBS) has not yet caught up with the new strategy and that new methods and tools are needed to help embed the new strategy's concepts and values in the routine operations of the department.

DynaRank, which is based on a hierarchical "scorecard" framework in Microsoft® Excel for the Macintosh and an IBM-compatible computer, ranks policy options by cost-effectiveness. Each ranking is a function of judgments about the relative importance of higher-level objectives and a variety of success criteria. Because DynaRank's hierarchical structure permits linking several levels of analysis, it can be used to integrate detailed analysis with high-level emphasis on components of defense strategy. It can also be used to intermingle subjective judgments about capabilities with other quantitative analyses of capabilities. Ultimately, DynaRank is intended not for technical-level operations research, but as an aid to high-level resource-allocation decisionmaking that is guided strongly by a sense of strategy. Figure S.1 shows a reduced "bottom-line" DynaRank scorecard dealing with the top level of defense strategy (components of shape, respond, and prepare now).

In Figure S.1, the rows of the scorecard are policy options, which can be either individual programs or program packages. The columns show different measures of the options' effectiveness, cost, and integration. There are columns for assessing an option's effectiveness for ensuring military capabilities for a broad diversity of scenarios and cases within scenarios, for environment shaping, and for preparing now for a variety of possible future challenges. There is a column showing a composite effectiveness, a column for the option's cost, and a column showing a measure of the option's cost-effectiveness. Further, DynaRank reorders the options so that they are in descending order of cost-effectiveness, i.e., the top option buys more for the money spent.

RANDMR996-S.1

Option (From Baseline)	Capabilities MTWs	SSCs	Environment Shaping	Strategic Adaptiveness	Net Effect	Cost	Cost Eff.
+ 20 B-2s							
+1500 SFWs							
+ Allied pkg							
+ 150 UAVs							
+ Smart Ship Technology							
+ 1 CVBG							
+ Homeport							
...							

MTWs = major theater wars.
SSCs = small-scale contingencies.
SFWs = sensor fuzed weapons.
UAVs = unmanned aerial vehicles.
CVBG = carrier battle group.

Each column can be an aggregate of subordinates.

Figure S.1—A Top-Level DynaRank Scorecard

The composite assessment, of course, depends on the relative emphasis given to the shape, respond, and prepare now components. Further, each of the individual evaluations depends on many assumptions, such as planning scenarios used to test capabilities, the perceived worth of forward presence for environment shaping, and the perceived worth of different types of hedges against strategic uncertainty. As a result, the DynaRank methodology must be both hierarchical and flexible. Indeed, as shown in Figure S.2, each of the top-level evaluations can be examined in more detail so that users can review and update evaluation criteria. Also, users can vary the relative emphasis across objectives. These changes can be made interactively in "real time," because DynaRank's simple spreadsheets run quickly and easily from a personal desktop or laptop computer with Excel.

This ability to examine and vary underlying assumptions is fundamental. Early efforts to develop decision-support tools based on multi-objective analysis methods often failed because they depended on a myriad of assumptions that could not be altered readily (or because issues were expressed in constructs more familiar to technical-level analysts than to policymakers).

A crucial feature of our approach is "uncertainty analysis." In practice, it is less useful to change assumptions one by one than to construct alternative "views," each of which involves a large number of assumptions. These views can correspond to different strategic perspectives that decisionmakers must consider. For high-level defense planning, one view might place a greater emphasis on environment shaping and preparing generally for uncertain futures, with less emphasis on near-term and midterm war-fighting capability. A second view might place a

RANDMR996-S.2

Option (From Base)	MTW Capability				Net Effect	Cost	Cost Eff.
	SWA		NEA	2 MTW			
	Case 1	Case 2					
+ 20 B-2s							
+1500 SFWs							
+ Allied pkg							
+ 150 UAVs							
+ Smart Ship Technology							
+ 1 CVBG							
+ Homeport							
...							

NEA = Northeast Asia.
SWA = Southwest Asia.

Each column can be an aggregate of subordinates.

Figure S.2—A Subordinate DynaRank Scorecard

greater emphasis on such war-fighting capability and associated readiness, sacrificing somewhat on the benefits of hedges against distant future threats that cannot currently be identified and on such optional environment shaping activities as peacekeeping operations. Clearly, each view would not only weigh each component of strategy differently from another (i.e., weight the columns in Figure S.1 differently), but would also entail different planning scenarios or assumptions within them. The approach we follow then ranks policy options under a variety of views so that it is easy to see whether some of them are robust across a reasonable range of views and which of them depend more sensitively on strategic judgments—and why. In practice, a number of options may be robustly cost-effective. They may, however, be infeasible because of political constraints. Therefore, DynaRank allows constraints to be applied when appropriate. Figure S.3 shows several options ranked by three different views. The shading indicates the ranked position of option across each view.

The methodology depends on whether the underlying evaluations in the scorecard are understandable and credible. In some cases, these evaluations can be traced to specific detailed analyses. Sometimes these analyses are embedded in DynaRank as low-level (high-resolution) spreadsheet models (e.g., a particular war-fighting scenario or a particular assessment of effectiveness for environment shaping). In some cases, the more detailed analyses are off-line (not part of DynaRank itself). And in some cases, more detailed analyses do not exist because decisionmakers and their staffs often draw on their general and specific knowledge and intuition to make judgments that are as credible as anything that could be generated by legions of analysts. For example, detailed analysis is unnecessary to convey that

RANDMR996-S.3

Portfolio "Views"			
Rank	MTW Emphasis	Shaping Emphasis	Adaptiveness Emphasis
1	**Allied pkg**	**Homeport**	**Boost ATBM**
2	**+1500 SFWs**	**Smart Ship Technology**	**+ 150 UAVs**
3	**Rapid SEAD**	**Allied pkg**	**Smart Ship Technology**
4	**+ 20 B-2s**	**New basing**	**Rapid SEAD**
5	**Smart Ship Technology**	**+ 1 CVBG**	**– 1 CVBG**
• • •			

ATBM = antiballistic missiles.
SEAD = suppression of enemy air defenses.

Figure S.3—Cost-Effectiveness Rankings of Options Representing Different Views Obtained with Different Emphasis, or Weighting, of Top-Level Measures

forces focused on long-range precision fires would not fare well in a long, dirty land war in jungles.

The DynaRank methodology is deliberately very flexible so that it can be applied by defense planners at a variety of levels, ranging from prioritization of a group of programs (e.g., tactical air force modernization) to evaluation of packages of programs involving high-level strategy

We believe that DynaRank fills a massive gap in existing methods and tools. As Figure S.4 suggests, its role is high-level integration of cost and effectiveness, and in some cases, of effectiveness across multiple objectives. It depends on a body of high-resolution analysis in many domains. Although it is an emerging tool that can be improved upon over time, it is currently ready for practical applications.

We are enthusiastic about the practical applicability of DynaRank. However, it has limitations that should be understood by the users of the methodology. "Additivity" of effects and costs, which are treated linearly by DynaRank, could be nonlinear, especially when the option effects are large, causing the cumulative results displayed by DynaRank to be meaningless. Measures are not likely to be independent, in which case weights on one measure imply a weight on another. Thus, the additive accumulation of weighted effects across measures, although cor-

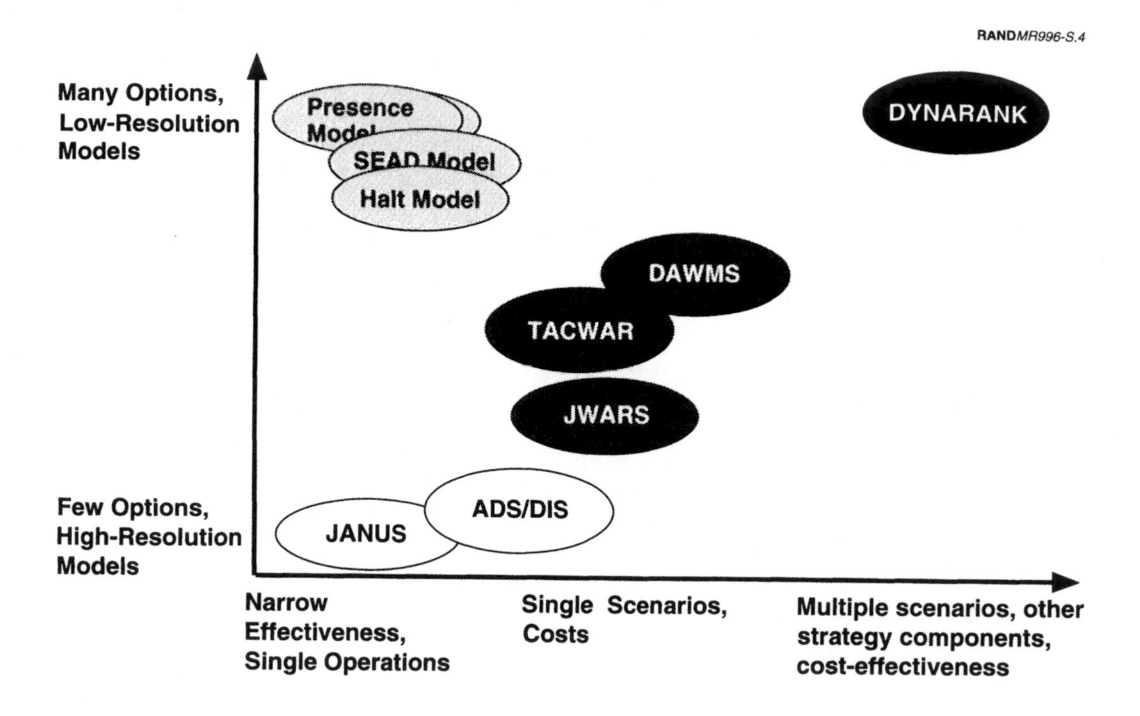

Figure S.4—DynaRank in the Spectrum of Defense Analysis Tools

rect, may reflect more weight than the user realizes on some of the measures. These are complicated mathematical issues, and the reader should refer to the good texts available on multi-attribute theory. We have generally been able to overcome these theoretical limitations through sensitivity analysis and some problem restructuring.

This report describes the development of DynaRank through a series of RAND policy studies, culminating in our application of it to the Quadrennial Defense Review (QDR). After an overview of the process and why it was developed, we briefly describe its application to a domestic transportation project and a long-range planning study for the Air Force. In Chapter Three, we go into more detail about its application late in the QDR process and describe how it assisted us in reaching certain conclusions.

The appendixes provide a detailed tutorial on the use of DynaRank and a listing of its menu functions.

Chapter One

INTRODUCTION

OBJECTIVE

This report describes a decision-support system called DynaRank, which is designed to assist the Department of Defense's (DoD's) development and updating of the U.S. defense program in a way that is explicitly consistent with the multiple objectives of the new strategy—shape, respond, and prepare now. We have been motivated by the observation that the current PPBS[1] has not yet caught up with the new strategy and that new methods and tools are needed to help embed the new strategy's concepts and values in the routine operations of the department.

BACKGROUND

Implementing a New Defense Strategy

The Quadrennial Defense Review or QDR (Cohen, 1997) defined a new defense strategy with three distinct components—shape, respond, and prepare now. The strategy seeks to elevate the importance of shaping the future environment and preparing for a wide range of possible future challenges by "transforming" U.S. forces for the next century's needs. The QDR also took a "capabilities analysis" perspective to the "respond" component by emphasizing the need to have operational capabilities for a highly diverse set of military contingencies and contingency circumstances, including circumstances such as responding to short-warning attacks and coping with the threat or use of weapons of mass destruction (both examples of exploiting U.S. Achilles' heels or what are sometimes called "asymmetric strategies"). Taken together these changes signaled a marked departure from previous strategy. They indicated the DoD's intention to base force planning decisions on a much broader intellectual construct than the two major regional contingency (MRC) "strategy" of earlier years. The capability to fight two simultaneous MRCs (now called major theater wars—MTWs) remains very important, but it is only one of many goals to be sought in developing the defense program and making choices within budgets.[2]

[1]The DoD Planning, Programming and Budgetary System.

[2]The two-MTW issue is also misunderstood in that the DoD is not attempting to maintain a force posture ready at all times—independent of other events in the world—to conduct two simultaneous

There are many challenges in making the new strategy meaningful. For example, the QDR calls for transforming the force. It draws on the ideas in Joint Vision 2010 and on technologies and concepts often discussed under the rubric of the revolution in military affairs (RMA); to pay for this, it calls for applying concepts of the revolution in business affairs (RBA). Indeed, paying for the transformation is a serious problem, since the current program is generally recognized to be underfunded. Further, it seems that the defense budget is more likely to remain constant, or even to shrink, than to grow in real terms. All of this, coupled with the inertial resistance to transformation that should be expected in large organizations despite enthusiasts for change, implies that it will take great skill to create and maintain a consistent program of modernization and transformation. History does not favor the QDR's hypothesis that force structure can be maintained and modernized by depending solely on cuts in infrastructure. Indeed, the QDR itself notes that while force structure has been cut by 33 percent, infrastructure has been cut by only 21 percent. This is not for lack of attention to infrastructure, but rather the result of many real-world constraints and resistances, such as congressional reluctance to close bases. A more vigorous attack on infrastructure costs is strongly warranted, but full success is unlikely and, as a result, the budget will be tight.

The problem, then, is to maintain a consistent defense program that addresses the three components of the new defense strategy and that nurtures a program of transformation to a force with future viability, and to do this in the face of budget pressures, changes in the external strategic environment, concern about protecting force structure, and internal DoD competition among the services. It is not clear that the current DoD PPBS is well suited for this type of planning and management. In particular, much of the PPBS process is bottom up, trade-offs are made within relatively narrow budget categories (the so-called stovepiping of defense planning), cuts are commonly allocated uniformly across categories ("salami slicing"), and the six-year planning horizon focuses most attention on the next few years. Procurement "bow waves" are often created by pushing acquisition costs out beyond the planning horizon to permit the PPBS plan to fit budget projections. The important issue here is how to revise the planning process so that the new defense strategy is more than rhetoric.

A Portfolio Approach

The need for new frameworks, methods, and tools for defense planning was in fact the motivating force behind the project discussed later in this document. A principal feature of our work has been a "portfolio-management" paradigm for defense planning (Davis, Gompert, and Kugler, 1996), which uses the same three-component approach as the QDR (Figure 1.1), but which in our analytical work

MTWs on short notice. Indeed, current forces are heavily engaged in shaping activities, including the Bosnia operation, some of which obviously draw down two-MTW readiness.

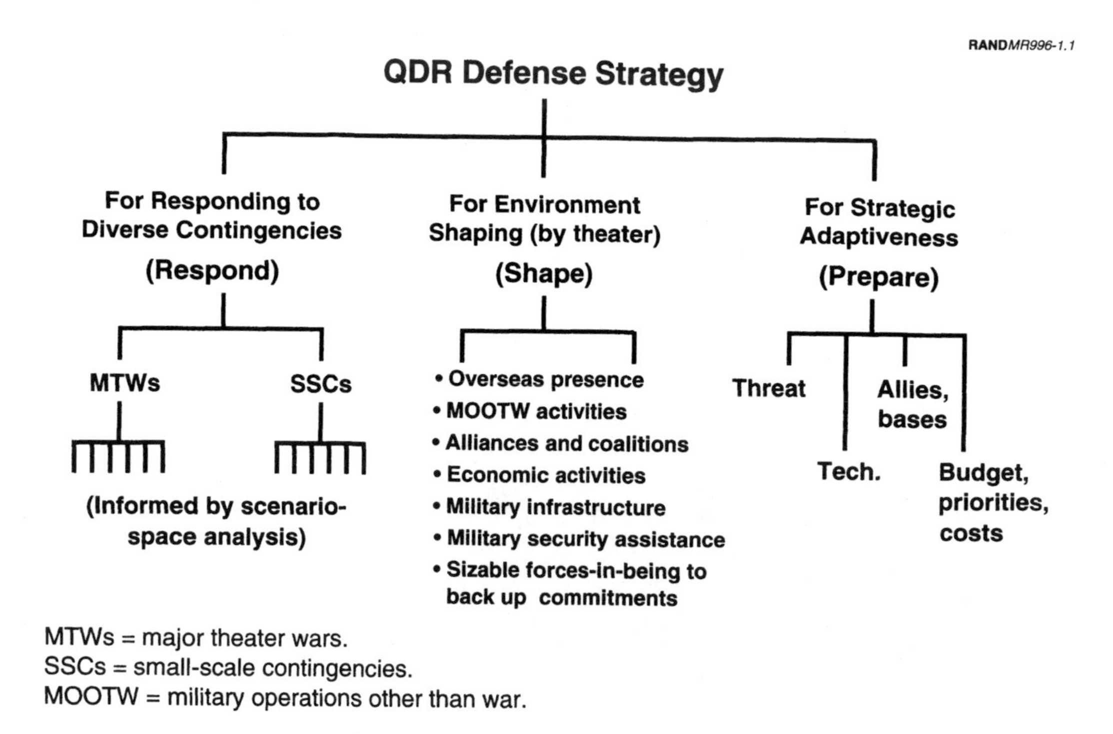

Figure 1. 1—Defense Strategy Components

translates into multi-objective analysis.[3] Our approach is essentially a new paradigm for defense planning modeled after the portfolio approach used by the financial world. Our vision of defense planning involves a robust portfolio of programs for research, force structure, modernization, infrastructure, and operations—one that explicitly addresses the various components of the strategy.

This portfolio should be continually evaluated with respect to risk and changing environment; it should be rebalanced over time as the emphasis on various components of the strategy changes.[4] This balancing and rebalancing can be aided by decision-support tools that clarify the logical implications for program priorities as one modifies the relative emphasis of different strategy components, of planning cases used to assess effectiveness in meeting objectives, and so on. Such tools should, of course, address both cost and effectiveness (from a variety of perspectives) and should allow users to apply constraints as necessary to reflect political or other imperatives.

---

[3]This work is reprinted, with slight changes, in the RAND collection *Strategic Appraisal 1997: Strategy and Defense Planning for the 21st Century* (Khalilzad and Ochmanek, 1997b). We refer to that volume as Strategic Appraisal 1997 in other citations.

[4]In some respects, this is what the DoD has done for decades, as "old hands" are apt to point out. However, the new strategy makes some of the key concepts such as hedging and adaptation explicit, and the approach we describe here is intended to make it easier to actually apply these concepts.

Consider the challenge to decision-support tools in more detail. Even if we focus on capabilities for contingencies (the respond component), and even if we consider a particular war scenario (e.g., protecting against another invasion by Iraq), there are typically multiple conflicting objectives in the use of military force. The relative value of these objectives changes as a function of operational strategy (e.g., restoring a border versus total defeat of an enemy), and the strategy is a function of the perceived enemy, our capabilities against that enemy and other political, strategic, and economic factors. There may be many scenario variations because of uncertainty about the conditions under which a conflict would play out. Warning time, enemy objectives, assumptions about allies, etc. fall into this category.

It is also necessary to evaluate the portfolio over different time periods. How well does the portfolio do today; how well does it (with planned modernization) play out against potential future threats? It is expected that the portfolio will be subjected to budget changes and that it will be necessary to rebalance the portfolio rationally in the face of such cuts. The defense options in the portfolio will be "apples and oranges" in the sense that infrastructure options may be traded for force structure options or modernization traded for force structure. Finally, the evaluation of the portfolio will necessarily involve a mix of quantitative and qualitative analysis, and a mixture of empirically rigorous data and subjective judgments. Models may be used to evaluate the portfolio in contingencies, but subjective judgments are necessary, for example, to estimate the value of various types and levels of presence on the shaping component of strategy.[5] In summary, the portfolio evaluation methodology needs to address the problem depicted in Figure 1.2.

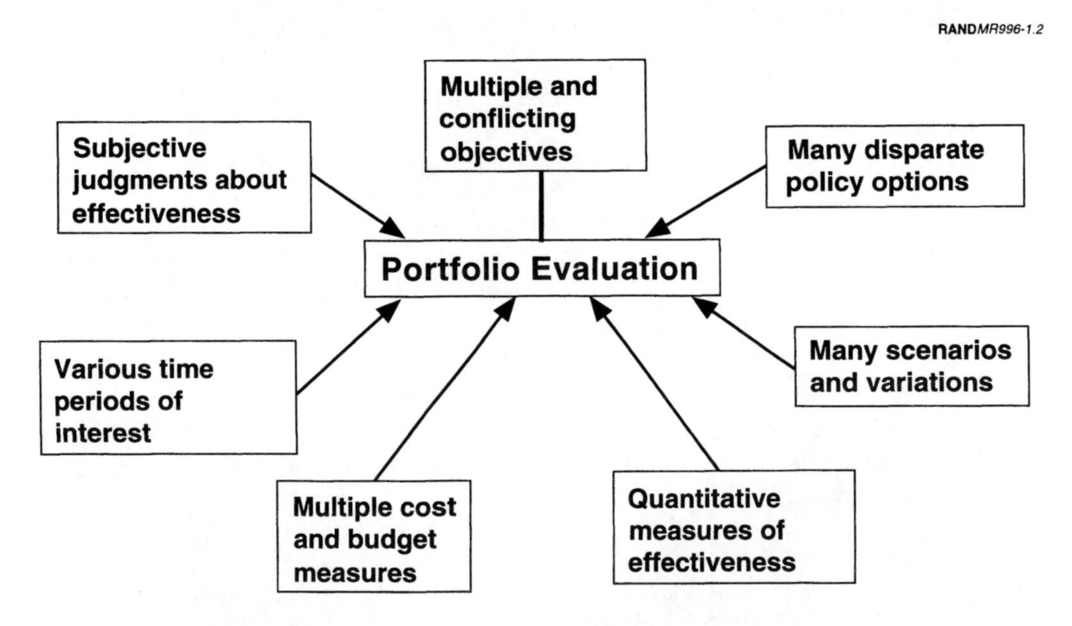

Figure 1.2—Dimensions in the Evaluation of a Defense Portfolio

[5]Of course the contrast here may be overdrawn because in reality most models include a large degree of judgment about parameter values and other factors.

The goals of strategic decision-support tools for defense strategy and portfolio management should be to (1) provide the ability to compare and evaluate a large set of dissimilar defense options across a broader set of cases, conditions, scenarios, and measures and (2) help to develop and maintain the rationale for the defense program and changes therein. This latter objective entails supplying a link to analysis; providing an understanding of how emphasis on strategy components, goals, cases, measures, etc., affect the program; helping to define a robust program that ranks high on cost-effectiveness regardless of the weights on components, and explaining why certain options (perhaps dropped from the program) do not rank high.

DYNARANK—A DYNAMIC, INTERACTIVE SCORECARD PROCESS

Against this background, then, this report is about a decision-support system (or tool) called DynaRank, which has been developed at RAND over several years in the course of performing planning studies on diverse subjects. Figure 1.3 illustrates schematically its components in evaluating defense program options in the new strategy. When reduced to its simplest form, DynaRank generates scorecard displays in which program options appear as rows and assessments of the options' likely value appear in columns with another column showing cost and a measure of relative cost-effectiveness.

This top-level view gives a bottom-line result. The DynaRank spreadsheets are hierarchical, so that, for example, the assessed value for MTWs of adding 20 B2s,

RANDMR996-1.3

	(Respond)		(Shape)	(Prepare Now)			
Option (From Baseline)	**Capabilities MTWs**	**SSCs**	**Environment Shaping**	**Strategic Adaptiveness**	**Net Effect**	**Cost**	**Cost Eff.**
+ 20 B-2s							
+1500 SFWs							
+ Allied pkg							
+ 150 UAVs							
+ Smart Ship Technology							
+ 1 CVBG							
+ Homeport							
...							

SFWs = sensor fuzed weapons.
UAVs = unmanned aerial vehicles.
CVBG = carrier battle group.

Each column can be an aggregate of subordinates.

Figure 1.3—Basic Elements of a Top-Level DynaRank Scorecard

which appears in the top left of the scorecard's body, is generated from subordinate spreadsheets considering a variety of different MTWs and cases for each. The evaluations for a given MTW case are generated by subordinate spreadsheet models in most instances, although in some the evaluation can be entered directly based on expert knowledge.

Weights are given to the various regional MTW scenarios, cases within the regional scenarios, and to objectives-related measures in the respond or contingency component of the strategy. Weights are also given to elements of the shaping component and to elements of cost such as acquisition and operations and maintenance (O&M). The latter may be used to differentiate costs by time period, to isolate operations and support (O&S) costs from investment costs, or to represent cost uncertainties. The result of this weighting is depicted as a ranked list of defense options. This ranking is commonly done by cost-effectiveness but could be done by cost or effectiveness alone. In fact, the DynaRank tool allows the "rollup" to net effect to be based on effectiveness minimums as well as weighted across the columns.

Obviously, the weights are subjective. They are also somewhat abstract. Who really "understands" viscerally what it means to weight one objective twice as heavily as another? It is therefore essential to understand how results depend on those weights, and to seek conclusions that are robust to substantial variations of the weights. Facilitating robustness is a major element of DynaRank and our related research program.

An important factor here is that DynaRank is a very fast desktop tool, which allows users to change assumptions and see results "right now." That is, it is a dynamic tool. In our approach, there is no attempt to seek a definitively "right" set of weights, but rather to explore how different assumptions and weightings affect the relative ranking of options (hence the name DynaRank). This process leads to multiple ranked lists that are in turn examined for commonalty and differences. Figure 1.4 suggests this idea schematically. Each of the "cards" on the left shows a list of options rank ordered by effectiveness under a set of assumptions. The purpose, as in most good analysis, is to explore results as a function of assumptions (and weights), and to then draw insights (right side of figure) that transcend the individual cases. An important part of this process is not only identifying robust options, but also developing an understanding, or rationale, for conclusions about robustness.

DYNARANK'S RELATIONSHIP TO OTHER MODELS AND METHODS

Defense planning requires a range of models and analytic methodologies. These vary from highly detailed and narrowly applied weapon-system models to theater-conflict models that aggregate great numbers of systems and are used to evaluate operational concepts or strategy in full-blown scenarios. Figure 1.5 illustrates this range of models and the position that the DynaRank approach occupies in the larger landscape. On the one hand, it can be considered an *integrative* methodol-

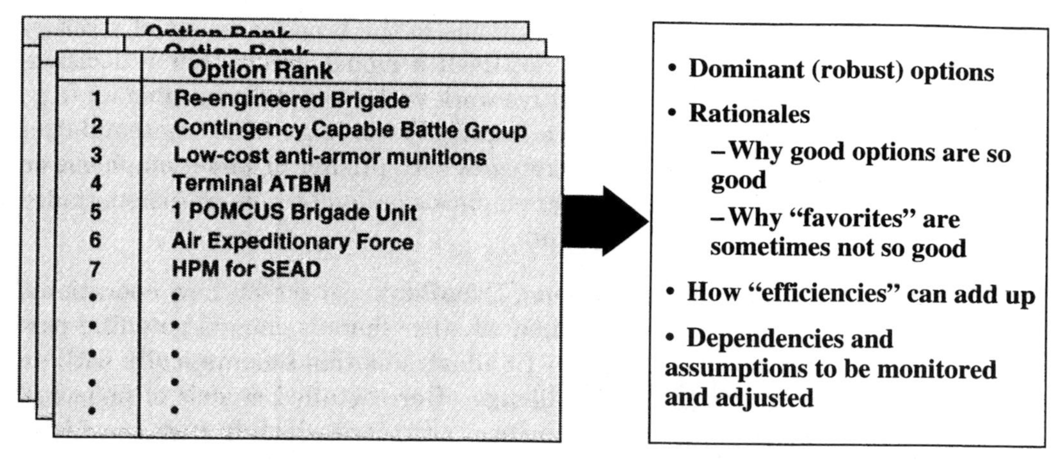

ATBM = antiballistic missile.
POMCUS = prepositioning of materiel configured to unit sets.
HPM = high-powered microwave.
SEAD = suppression of enemy air defenses.

Figure 1.4—General Conclusions Based on Interactive Ranking

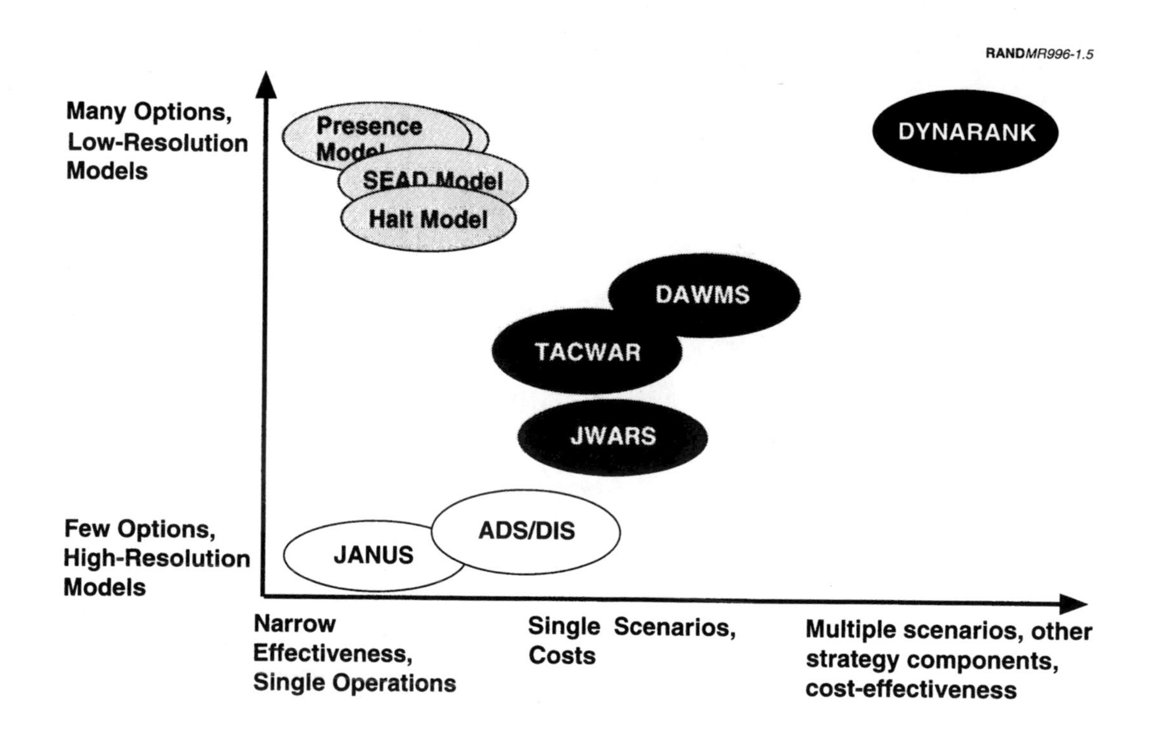

Figure 1.5—The Spectrum of Defense Analysis Models and Methodologies

ogy that takes results from models at various levels and compares outcomes across a broad set of objectives, cases, measures, and costs. On the other hand, because of its hierarchical structure, it can use embedded models to provide a linkage between the detailed representation of systems to the broader issues of strategy and defense planning. DynaRank is not itself a model, but rather a decision-support tool. It can be used for integrative work within a particular domain (e.g., tactical air force modernization) or at a strategic level (e.g., when contemplating packages of options corresponding to strategies with greater or lesser emphasis on forward deployment, or greater or lesser emphasis on long-range precision-guided weapons rather than forces on the ground).

When used in its most aggregate version, DynaRank can screen new operational concepts and potential capabilities to show whether there is enough potential pay-off to justify in-depth analysis. Figure 1.6 illustrates this schematically with an example focused on the halt-phase challenge. Here detailed models of phases of operations are linked directly to a contingency scorecard which in turn provides a weighted aggregation or rollup of scenario cases to a strategic-level scorecard. Similarly, models representing various aspects of military presence feed an environment-shaping scorecard that is also rolled up to the strategic scorecard. This linkage permits the top-level view of a defense program to have a tight and auditable coupling to analysis in depth.

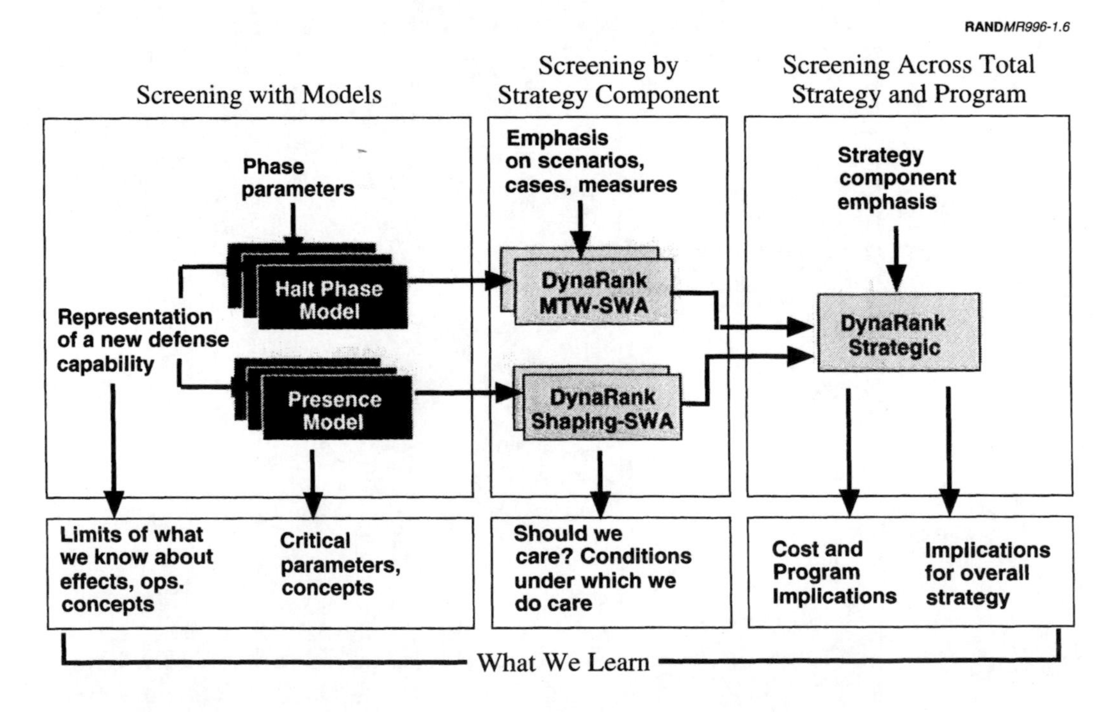

Figure 1.6—DynaRank Linked to Models and Used to Screen New Capabilities at Several Levels

ORGANIZATION OF THIS DOCUMENT

The next chapter provides an overview of the features of the DynaRank decision-support tool, its development history, the essential components, and the general process of using it. Chapter Three describes the application of DynaRank to the evaluation of options during our work with QDR defense policy options, briefly discussing the kinds of conclusions we reached regarding priorities in defense planning for the new strategy. The appendixes of this document provide a tutorial that should enable a new user to set up and apply DynaRank and provide a step-by-step example of its use, and list DynaRank's functions.

Chapter Two

DYNARANK—AN OVERVIEW

HISTORY OF DEVELOPMENT AND APPLICATIONS

Why History Is Relevant

Because a decision-support system for applying a multi-objective strategy such as shape, respond, and prepare now necessarily implies methods that have generally not been used by DoD decisionmakers (for a variety of good reasons related to complexity and the fuzziness of many analyses reflecting subjective judgments explicitly), it is useful to provide some background on DynaRank's development and to emphasize that the methods it incorporates have in fact been used successfully in the past. They are not merely "academic curiosities," but practical techniques. To use them well requires sophistication, to be sure, but DoD's planning has often been quite sophisticated over the decades.

Scorecards Used in Policy Analysis

DynaRank is a systematic application of "scorecard" methodology. Historically, the scorecard, sometimes called a stoplight chart, has been an important method of presenting the results of a policy analysis to decisionmakers. Colors (or patterns and shadings in black-and-white depictions) represent the relative value or contribution of a policy alternative to each of a variety of measures. The two-dimensional display permits a quick view of how all of the policy options fare across the measures of interest. This is particularly important because good policy analysis recognizes that policymakers must bring to bear a number of value judgments and constraints that cannot and probably should not be buried in technical analyses done by their staffs. That is, there should be a separation between what can be accomplished "technically" and what must be assessed by the decisionmakers themselves. The scorecard approach permits this. For example, some of the columns may show "technical" assessments, others may show various and sundry subjective assessments by interested parties and another may show cost. The decisionmaker can then draw conclusions based on an integrated variety of information.

In the early 1970s RAND used scorecards to show the value of alternatives in an award-winning policy analysis of Dutch water management that had important effects on subsequent Dutch programs (Goeller et al., 1977 and 1985). These scorecards were generally static vugraphs showing the culmination of a set of research. Since that time, scorecard methods have become ubiquitous and even appear in popular media such as *Consumer Reports*. While it is easy to nitpick them, or to carp about how much is assumed in constructing them, they have proven themselves over time. Their value, however, depends on the quality and integrity of those who develop and use them.

Transportation Planning—The FORWARD Project

In 1993, RAND initiated a study for the Dutch government that reviewed its freight transportation policy for the next 15–20 years. The policy issues were important because of the Dutch interest in becoming a transportation hub. The port of Rotterdam and the Netherlands' central location in Europe made transportation an attractive focus for economic development. At the same time, the movement of large amounts of freight and passengers through the Dutch system of waterways, rail links, and highways was forecast to create enormous problems for the environment and would dramatically increase congestion on already overloaded highways. The FORWARD (Freight Options for Road, Water, and Rail for the Dutch) project (Hillestad et al., 1996) examined some 200 policy alternatives for mitigating the negative effects while maintaining the economic viability of the Netherlands as a transportation hub. The policy options varied from building new transportation infrastructure (building a dedicated freight rail system from Rotterdam to Germany) to changing regulations (removing some weight limits on trucks), to technological fixes (building cleaner diesel engines). The measures of interest were emissions (broken down by type of emission), noise (as it affected populated areas along a highway), safety (accidents, hazardous spills, and fatalities), and a range of economic effects such as employment and added value to the Dutch economy. Thus, the Dutch were faced with "apples and oranges," much as the Department of Defense is when it considers a strategy of shape, respond, and prepare now.

The FORWARD model (Carrillo and Hillestad, 1996) was built to evaluate these options and, as part of the development, one of the displays of this model was a scorecard of values and colors showing how well each option faired on each of the measures and on cost. Because the model was implemented in a spreadsheet, an analyst or policymaker could change the parameters of the model and see the impact on the "scores." The ability to weight the measures was included as well as the capability to rank the policy options on cost-effectiveness. In addition, cumulative plots of cost and aggregate effectiveness were created to approximate the addition of effects and costs of the most promising options. One could then dynamically change weights on measures, model parameters, or costs and observe how the options ranked and how the cumulative effects behaved. From this, RAND was able to develop general conclusions about the most robust options, the order in which policy alternatives ought to be considered, and how the alternatives depended on the weighting of measures. Surprisingly, the set of conclusions reached was quite different from the original hypotheses because the addition of a number of options with small effectiveness improvements but low costs (and sometimes cost savings) added up to relatively large gains in effectiveness at very little cost when the cumulative effects were considered. These "efficiency options" then became an important focus of Dutch long-range planning.[1] Figure 2.1 illustrates the FORWARD system model and scorecard.

It was as a result of this work that we observed and realized the power of *dynamic* scorecards, linked to models and provided with modern and ubiquitous computational tools to do ranking and plotting, to enhance policy analysis in general, especially when there were multiple policy alternatives and multiple and sometimes conflicting measures of those options.

An Application for the Air Force

During fiscal year 1996, RAND's Project Air Force performed a study for the Air Force's Office of Long Range Plans to study Air Force capability, force structure, and cost-saving options for the future Air Force.[2] The study's objective was to evaluate a broad range of these options and packages of these options against an uncertain future defined by many scenarios of potential conflict, including variations that reflected different assumptions about enemy capabilities and strategy and that took place in different parts of the world. We used the same general approach as the FORWARD study described above, but the models represented conflict and the measures were such dimensions of success and cost as enemy penetration, attrition, time to friendly force success, etc. These conflict models fed results to a large scorecard. Again, we added tools to rank options and display cumulative effects. Figure 2.2 illustrates the content of the MRC scorecards.

With this scorecard we could change model parameters or weights on scenarios and measures and look at the influence on the cost-effective ranking of options. The set of conclusions reached in this case was quite different in kind from those reached in the FORWARD study. In many Air Force shaping scenarios, we could not differentiate the effects of options because the base case was "too easy" for the Blue side. Marginal changes in capability simply did not have much effect because baseline capabilities were already adequate. In addition, when we focused on the scenarios and examined some tougher cases, we were surprised to find that again many of the options had little effect—the cases had been made difficult in ways that the options did not help, for reasons more evident in retrospect than in advance. This, in turn, caused us to identify new options and packages that would have an effect and brought us to the realization that the problem in most current contingencies was not getting more force into the theater or making marginal improvements in capability, but rather defeating the asymmetric strategies of an opponent. Asymetric strategies included fighting in urban areas and denying airfields and ports. Such things as tactical missile defense to defend airfields, getting

---

[1]See, for example, the article in *OR/MS Today*, "Dutch Move 'FORWARD' with OR," June 1996.

[2]The study was led by Natalie Crawford. A summary of conclusions from it and a somewhat earlier study for the Office of the Secretary of Defense and the Joint Staff is available in Davis, Hillestad, and Crawford (1997).

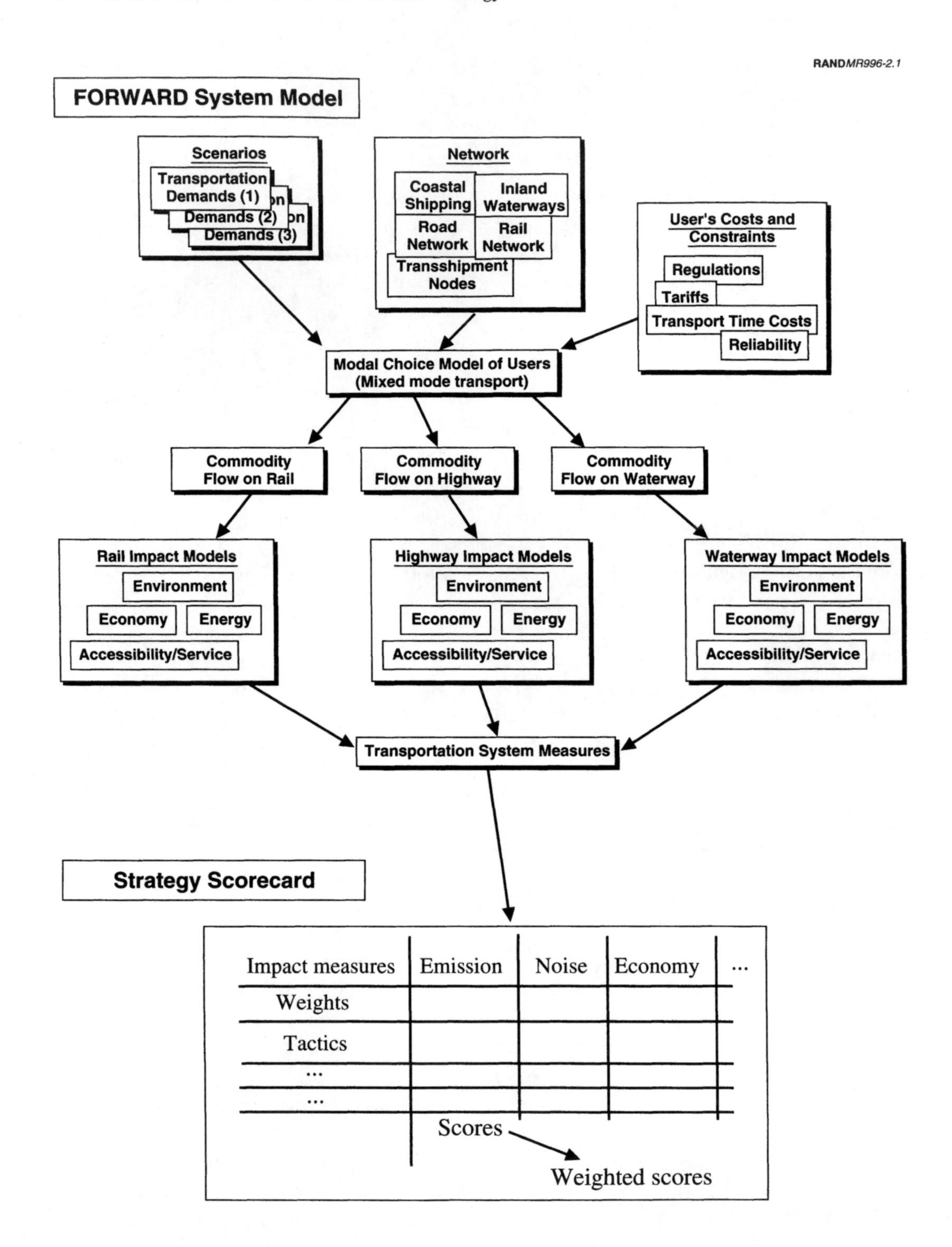

Figure 2.1—The FORWARD Model and Scorecard

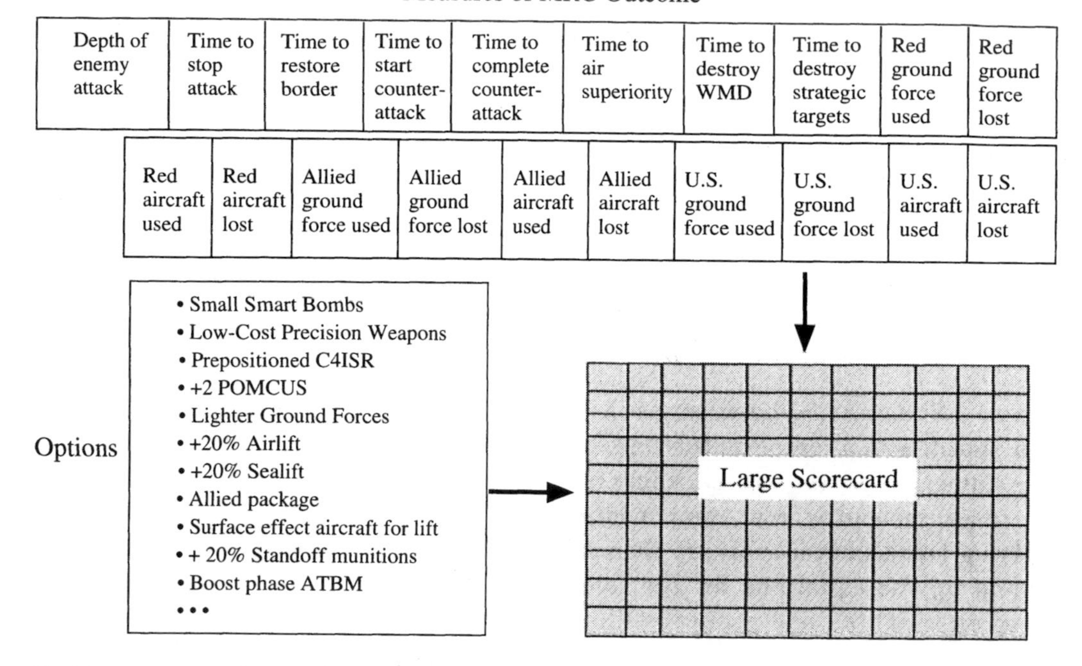

WMD = weapons of mass destruction.
C4ISR = command, control, communications, computers, intelligence, surveillance, and reconnaissance.

Figure 2.2—Schematic of Measures and Options Included in the Air Force Shaping Scorecard

allies ready to fight delaying actions while we brought in forces, and developing systems that could be used in urban fighting to reduce collateral damage and civilian casualties as well as identify the bad guys were most important.

Strategic-Level Application in Support of the QDR

During fiscal year 1997, we generalized the scorecard approach and applied it in support of the Quadrennial Defense Review (Davis, Kugler, and Hillestad, 1997). The generalization permitted rapid construction of new, hierarchical scorecards through the DynaRank templating process. All equations for applying weights to compute aggregate performance were created automatically. The structure of the scorecard in terms of measures and policy options, base-case representation, and goals was formalized. In addition, a number of tools were created to perform cost-effectiveness calculations, plot results, rank options, color code the scorecard, link results to scorecards in a hierarchy, and help isolate "robust" options (those that remain high in a ranking as weights on measures, cases, and scenarios are varied). With this flexibility, we were able to evaluate a set of strategic-level QDR defense options across a hierarchy, ranging from detailed scenario analysis to strategic evaluation with respect to the shape, respond, and prepare now aspects of the new defense strategy. We could construct different "views" of the options in which different components and subcomponents of the strategy were given more or less weight to correspond with various strategies. These different views allowed us to identify defense options that were somewhat independent of the emphasis placed on the strategy component. We could also let cost components reflect cost uncertainty or, alternatively, costs by period to determine the influence of such uncertainties or the period of time being emphasized. Chapter Three describes the QDR application in more detail.

OVERVIEW OF DYNARANK METHODOLOGY

Steps in the Process of Applying DynaRank

Figure 2.3 shows the basic steps involved in applying DynaRank. Initially the user specifies in a template the framework of his or her analysis. That is, the user lists names of options and packages to be evaluated, strategy components, scenarios, cases, measures, cost components, and hierarchy of representation. The underlying DynaRank machinery then creates the scorecard based on this template, setting up the equations for the calculation of aggregate performance and cost based on weights, and generally filling out all of the scorecard except for the actual performance values. The second step is to fill in the cost and performance squares of the scorecard. This might involve recording results from contingency models as shown in Figure 2.4 or entering subjective judgments. Other alternatives include providing the values from subordinate scorecards (which themselves might be filled in from models or subjective judgment), or to actually insert equations or "production functions" for computing the values within the cells. Once the scorecard is completed, the DynaRank process provides interactive features that permit the analyst to vary the weights on strategy components, scenarios, cases, measures, and cost components and to rollup the values to an overall performance

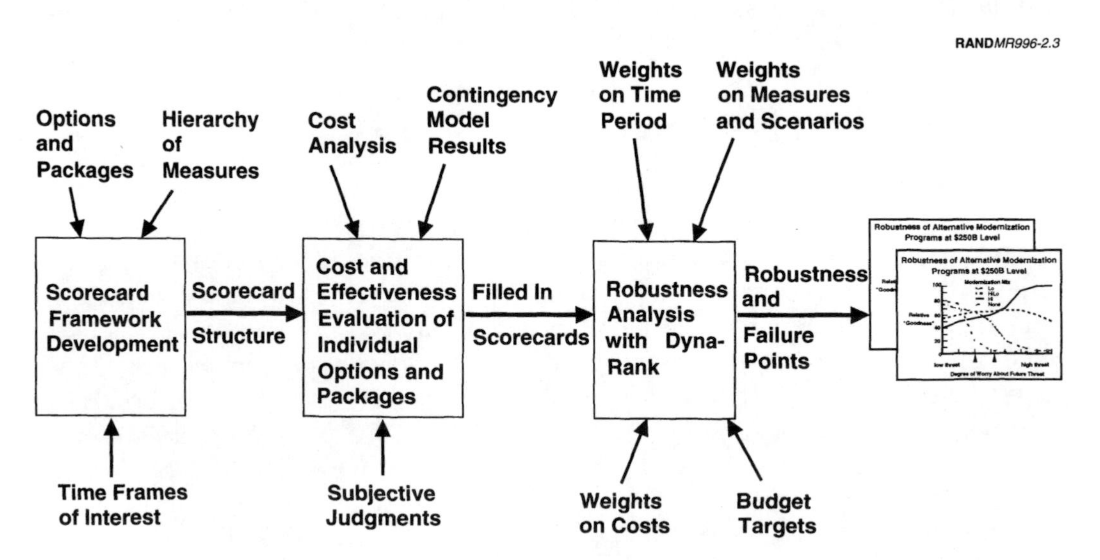

Figure 2.3—The Process of Applying DynaRank to Evaluating Policy Options

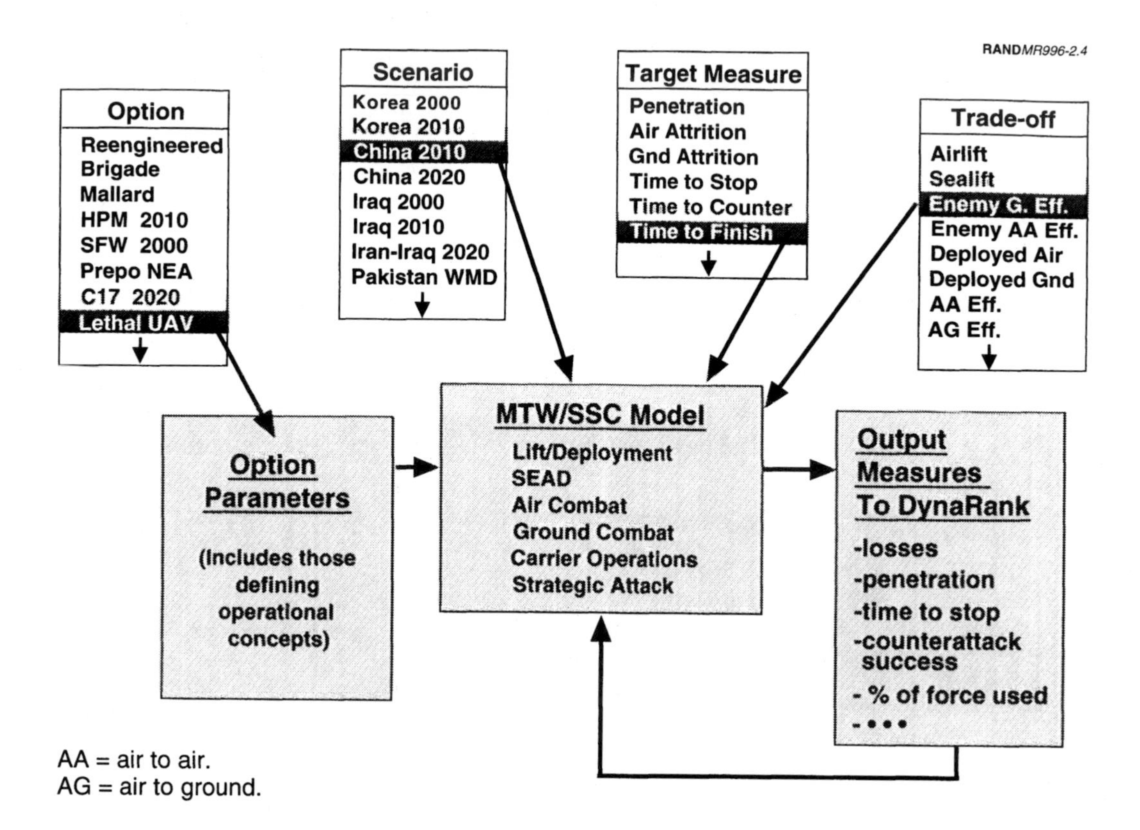

Figure 2.4—Model Outputs As a Source of Input to DynaRank

and cost evaluation of the option. This rollup is used to determine rankings of options, as well as displays of accumulative value. Additional tools within DynaRank permit the assessment of the robustness and commonalty of options across "views" or rankings of options dependent on particular weights on performance measures.

The Features of DynaRank

Basic Structure of a Scorecard. Figure 2.5 illustrates a DynaRank scorecard. Basically, the rows of the scorecard represent the policy alternatives or options, and the columns represent the criteria against which the options are to be measured. The intersection of the rows and columns or body of the scorecard contains the evaluation of each option against each measure. In our scorecard structure, we assume that this evaluation is normalized so that the scale goes from 0–1, 0–10, or 0–100. This can be set by the user. Generally there is a base-case row (e.g., the case corresponding to the current defense program). Most subsequent rows show possible changes on the margin and their consequences for cost and effectiveness. The color scale can be defined by the user. The goal row (if included) indicates the desired performance on each measure. The measures of effectiveness are represented hierarchically with three levels allowed in a single scorecard. The weights of each measure appear in the box with the name of the measure. These weights

RANDMR996-2.5

Color Blank Cells

High Color Value: 100

Low Color Value: 0

CE Cost Wt.: 1

Strategic Level Scorecard

					Contingency Capability (RESPOND) 1			Environment Shaping (SHAPE) 1		Strategic Adaptiveness (PREPARE NOW) 1		Wt or Min ↓ Wt	Costs	
			Top Level Measures ->		Contingency Capability (RESPOND) 1			Environment Shaping (SHAPE) 1		Strategic Adaptiveness (PREPARE NOW) 1		Wt	Wt Annualized Cost 1	
			Mid Level Measures ->		SWA 1		2 MTWs or MTW& MOOTW 1	Overseas Presence 1	Security Assistance 1	Transforming the Force 1	Other Hedges 1	Wt		
			Base Level Measures ->		Case 1 1	Case 2 1						Wt		
			Goal -->		100	100	100	100	100	100	100			
Graph			Option	Flag.	Base case row							Aggregate Column		Total Cost
					65.9	40.0	75.0	65.0	65.0	65.0	65.0	64.7	0	0
1	1	Modernization	Arsenal ship	1	71.3	40.0	79.0	67.0	65.0	70.4	65.0	67.0	0.251	0.251
1	2	Modernization	ATBM system	1	65.9	42.5	85.0	80.0	65.0	65.0	65.0	69.0	1	1
1	3	Modernization	F-22s anti-armor	1	67.1	40.0	82.0	65.0	65.0	66.2	65.0	66.1	0.05	0.05
1	4	Modernization	C4ISR package	1	84.8	42.5	82.0	65.0	65.0	80.0	65.0	70.1	0.4	0.4
1	5	Modernization	Rapid SEAD (HPM)	1	75.2	40.0	85.0	65.0	65.0	74.3	65.0	68.7	0.11	0.11
1	6	Modernization	Standoff munitions	1	73.7	40.0	80.0	65.0	65.0	72.8	65.0	67.4	0.2	0.2
1	7	Efficiency	Allied pkg	1	96.2	94.2	95.0	75.0	75.0	80.0	65.0	80.0	0.3	0.3
1	8	Efficiency	Faster tacair deployment	1	68.0	40.0	79.0	65.0	65.0	67.2	65.0	65.9	0.105	0.105
1	9	Efficiency	Double surge sortie rates	1	72.4	40.0	82.0	65.0	65.0	71.5	65.0	67.5	0.05	0.05
1	10	Force Structure	Minus 2 TFWs	1	65.9	40.0	62.0	61.0	65.0	65.0	65.0	61.8	-0.7	-0.7

Figure 2.5—A DynaRank Scorecard (See Plate I for color illustration)

are initialized to 1.0 for each measure. The aggregate-value column represents the product of the normalized weights on each measure multiplied by the corresponding effectiveness and summed. Weights are normalized when used so that weights of 1.0, 2.0, and 1.0 across a set of three measures would actually be transformed to 0.25, 0.5, and 0.25 when used to multiply the option values. The combined weight for any column is the product of all the normalized weights in its hierarchy. The cost columns (only one is shown in the scorecard in Figure 2.5) are assumed to represent different components of cost, while the aggregate-cost column represents the sum of the weighted cost components for each option. For example, one might show a cost for the current five-year defense plan (FYDP) and an eventual annual steady-state cost. The user might consider one or another of these more important. The cost weights are not normalized so that cost components can be summed directly; or, alternatively, the weights can be used as discount factors when the components represent costs within different time periods.

Creating a Scorecard. To go through these items in more detail, we'll illustrate the creation of a scorecard. To create a DynaRank scorecard the user works with a DynaRank template to define the options, hierarchy of measures, cost components, and value scale for the scorecard. Figure 2.6 illustrates a template for the scorecard in Figure 2.5. The options (and option categories if desired to classify the options) are entered as a list, along with the measures and the cost components. Once the user is satisfied with the list, the Create Scorecard function is used to produce the scorecard structure, without performance values. This function permits the rapid structuring and restructuring of the basic policy problem and its criteria. The Create Scorecard function takes care of not only the tedium of producing and modifying scorecards, but it also places equations in the scorecard cells that enable automatic rolling up of performance and cost as well calculations of cost-effectiveness. The process also permits naming of scorecards and builds a catalog of available scorecards and templates in the workbook. Templates and scorecards in DynaRank carry hidden data that identify the worksheet type type so that the user cannot inadvertently perform a scorecard function on a template, etc.

Coloring the Scorecard, Varying Weights, Ranking, and Plotting Results. Perhaps the simplest function to be performed is the coloring of the scorecard body to provide a two-dimensional picture of the relative score of options on meeting the individual measures. The Color and UnColor functions in the DynaRank menu cause this to happen. The colors range from bright red, representing performance near zero, to bright green, representing performance close to the goal. There are five colors representing performance spaced evenly between the high and low color values specified by the user for the scorecard. Yellow is the midrange color. The colors from worst performance to best performance are then red, orange, yellow, chartreuse, and green.

Aggregation of performance occurs through the use of weights or multipliers set by the user. There are several aggregated views available. One view, available in the main scorecard, is the aggregate column. It is created by rolling up all of the criteria columns, which is done by multiplying by the appropriate weights. The

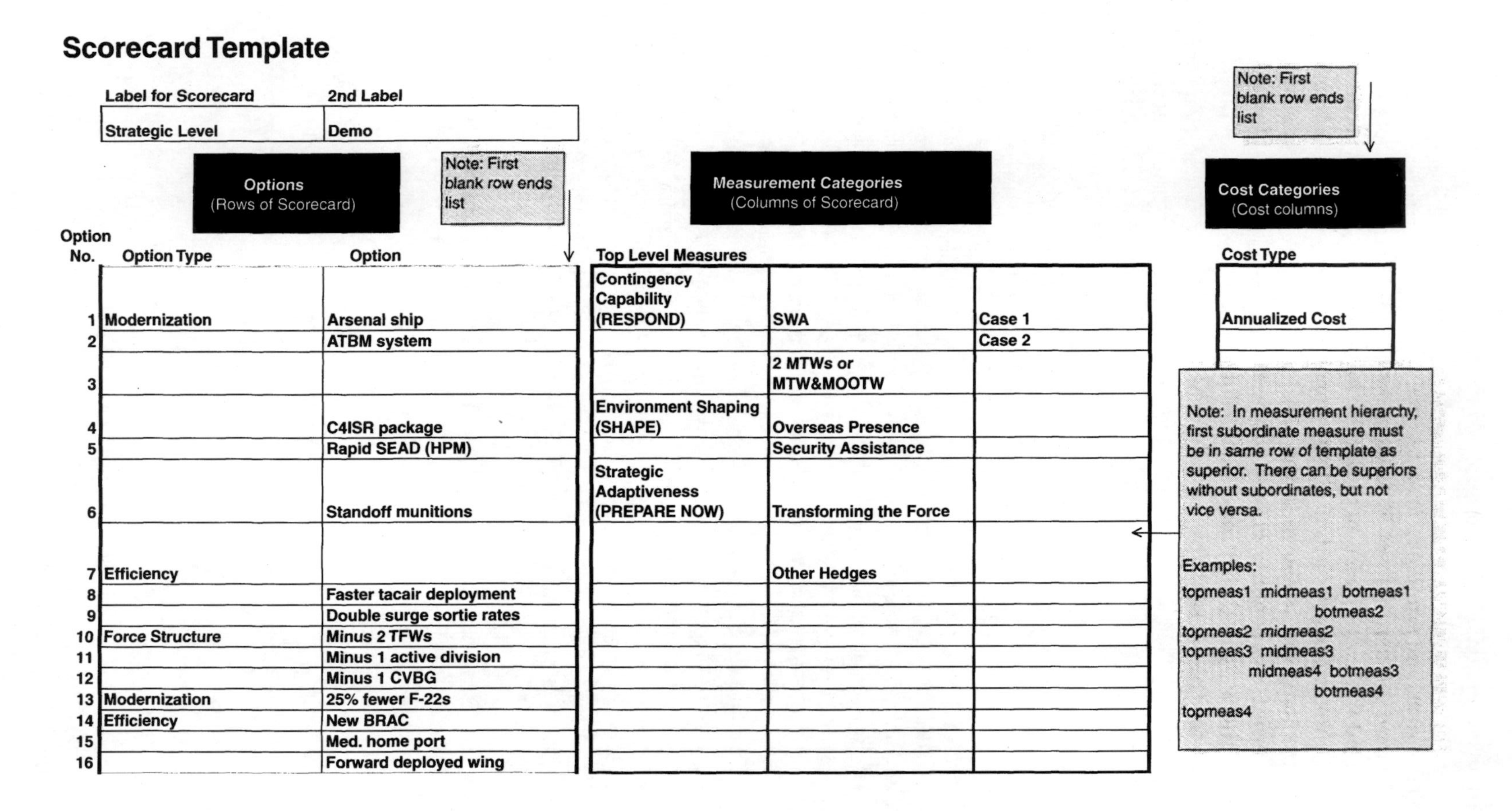

Figure 2.6—A DynaRank Template

DynaRank scorecard can be created with a hierarchy of measures; top-level (primary) measure, level 2 measures (secondary), and level 3 measures (tertiary). There are also subsidiary scorecards created on the same worksheet that show the partial aggregation of results to primary measures, and to the secondary measures. These scorecards are colored at the same time as the main scorecard. As an alternative to weighted aggregation at any level, the user can select the rollup-to represent the worst case. By changing the name Wt in the aggregate column to Min, the worst case at that level will be used by DynaRank. In addition, weights can be used at one level and the worst case at another level. For example, the user might evaluate contingency capability for a given region with a worst-case scenario, but capability in other regions might be measured as weighted sums of cases. That is, if there are two Southwest Asia (SWA) contingencies, the user could use as the aggregate the contingency with the worst outcome rather than the weighted sum of the two. Overall strategic effectiveness could then be calculated as a weighted sum across the strategy components.

The aggregate-performance column and the aggregate-cost column help to compute cost-effectiveness, provide cumulative costs and effectiveness, rank options, and display ranked lists. Figure 2.7 illustrates resulting plots of options by cost-effectiveness. Options can be ranked by cost, effectiveness, or cost-effectiveness. The figure shows how the performance increases as more and more options are added in rank order and how the cost builds as these options are added as well. As a matter of theory, the aggregate-performance chart can be misleading, because in the general case the options being considered may interact with one another so that the cumulative effectiveness is not the sum of marginal contributions. For example, if one has several options for accomplishing a goal, it may be in the real

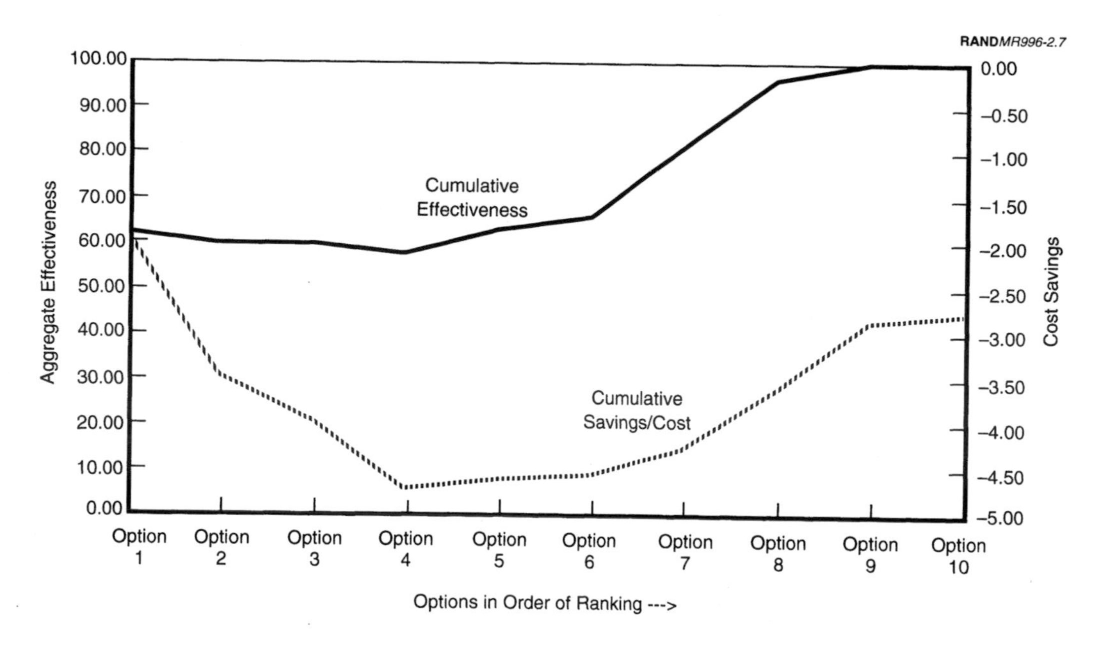

Figure 2.7—Cumulative Cost-Effectiveness and Plots of Ranked Options

world that completing the first option solves the problem and makes the second unnecessary. In the spreadsheet, however, both would contribute independently (a linear approximation). Despite the potentially treacherous nature of this column, we include it because, in practice, the linear approximation is often rather good in marginal analysis. Also, the user aware of the issue can often redefine options (e.g., combine options sensibly) to increase their independence and make the linearity assumption valid. Or the user may choose to edit the evaluation algorithms to account for the nonlinearity.

Accumulating Lists Representing Different Views. After options are ranked, it is often desirable to compare these with another ranked list of the same options but obtained with different weights or worst-case (Min) criteria. DynaRank provides the Rank sheet and the Results sheet for this purpose. After a ranking is performed, the user names the sheet on which the rank is to be saved and gives the "view" a name. This is transferred to the appropriate sheet for later use in assessing robustness, etc. There are additional DynaRank functions that are tailored for these sheets. For example, the Color Rank Results function causes common options in the top elements of the list to be colored the same. Figure 2.8 illus-

RANDMR996-2.8

View A	View B	View C
Emphasizes Contingencies	Emphasizes Shaping	Emphasizes Cost
New BRAC	25% fewer F-22s	Minus 1 active division
Double surge sortie rates	New BRAC	Minus 1 CVBG
Allied defense package (helos/ATACMS/ advisors)	Allied defense package (helos/ATACMS/ advisors)	25% fewer F-22s
Equip F-22s for anti-armor missions	Forward deployed wing	Minus 2 TFWs
Rapid SEAD (HPM)	Med. home port	New BRAC
Forward deployed wing	Arsenal ship plus allied package	Forward deployed wing

Figure 2.8—Alternative "Views" Stored in a Rank Sheet with Common Items Colored Alike
(See Plate II for color illustration)

trates this capability, which helps to quickly identify the common or robust options at the top of the list. In the Results sheet, the user is allowed to weight the views and obtain the best composite ranking. As long as the same set of options is being evaluated, the underlying models can also be different for the various views; either parametrically, structurally, or both.

Other DynaRank Features. DynaRank is capable of linking scorecards and creating hierarchies of scorecards, with the aggregate results of one feeding another. For example, the strategic-level scorecard of Figure 2.5 obtains its values from more detailed SWA scorecards, shape scorecards, and prepare scorecards. Figure 2.9 shows the underlying SWA scorecard. This scorecard, in turn, is linked directly to models of the conflict that it represents. This process of embedding can go on indefinitely and permits an auditable link between representations and criteria at different levels. A number of "combine" functions are available in DynaRank to support this. It is also possible to show the colors of one scorecard superimposed on the values and colors of another. This "transfer" feature can be used, for example, to show the risk of an option, evaluated on a separate scorecard, to be displayed along with the option value on another scorecard.

These features and others are described in more detail in Appendix A.

RANDMR996-2.9

☐ Color Blank Cells

High Color Value: 100

Low Color Value: 0

CE Cost Wt.: 1

SWA Late Scorecard

					Top Level Measures ->	SWA (1)						Costs	
					Mid Level Measures ->	Case 1 (1)		Case 2 (1)					
					Base Level Measures ->	Enemy Penetration at Halt (1)	Defeating and Occupying Iraq (1)	Enemy Penetration at Halt (1)	Defeating and Occupying Iraq (1)	Wt or Min		Annualized Cost (1)	
					Goal -->	100	100	100	100				
Graph	No.	Option Type	Option	Flag.		Base case row				Aggregate Column			Total Cost
						41.8	90.0	0.0	80.0	52.9		0.0	0.0
1	1	Moderniza-tion	Arsenal ship	1		52.6	90.0	0.0	80.0	55.6		0.3	0.3
1	2	Moderniza-tion	ATBM system	1		41.8	90.0	0.0	85.0	54.2		1.0	1.0
1	3	Moderniza-tion	Equip F-22s for anti-armor	1		44.2	90.0	0.0	80.0	53.5		0.1	0.1
1	4	Moderniza-tion	C4ISR package	1		74.5	95.0	0.0	85.0	63.6		0.4	0.4
1	5	Moderniza-tion	Rapid SEAD (HPM)	1		60.4	90.0	0.0	80.0	57.6		0.1	0.1
1	6	Moderniza-tion	Standoff munitions	1		57.4	90.0	0.0	80.0	56.9		0.2	0.2
1	7	Efficiency	Allied defense package	1		97.3	95.0	83.4	85.0	90.2		0.3	0.3
1	8	Efficiency	Faster tacair deploy-ment	1		46.1	90.0	0.0	80.0	54.0		0.1	0.1

Figure 2.9—A Subordinate Scorecard (This provides results used by the Strategic Scorecard of Figure 2.5) (See Plate III for color illustration)

Chapter Three

APPLICATION OF DYNARANK STRATEGIC PLANNING IN THE QDR

OBJECTIVES OF STRATEGIC ASSESSMENTS FOR THE QDR

As part of RAND's work in support of the Quadrennial Defense Review, we applied the DynaRank process to a broad range of capability- and budget-oriented options within the QDR. While this came too late to affect the QDR itself, it documented and confirmed preliminary analysis that we had provided earlier and offered a testing ground for DynaRank. Below we describe the application of DynaRank to that work.

One important issue for us was to examine how a variety of options would be ranked or rated as a function of how seriously one considered the various components of the emerging strategy of shape, respond, and prepare now. For example, what options would look good only for contingency capability (respond) and what options would contribute to shape and prepare now as well? How would heavy emphasis on major regional contingencies affect the defense program's ability to satisfy the other components of strategy? If we remain focused on the current period of time, what vulnerabilities will that lead to with respect to long-term adaptiveness of the defense program to future enemies and capabilities?

There was also concern about the budgetary impact of the new components of defense strategy. What reductions in forces might be made to permit modernization within a budget that might be generally declining in real terms, or that might be constant in real terms but plagued with the consequences of underfunding in the FY98 program? What is the impact of emphasizing presence operations relative to buying capabilities to fight major theater wars? Above all, we wanted an understanding of whether there were defense options that were robust in the sense that they remained cost-effective across a range of emphasis on the strategy components and across a range of emphasis on the scenarios, cases, regions, and performance measures for the second and third tier of defense criteria.

This chapter describes how we applied DynaRank to an investigation of these QDR issues to further illustrate the setup and application of DynaRank to defense planning problems.

DEFINING THE HIERARCHY OF DEFENSE CRITERIA

The Respond Component of Strategy

This dimension involves the ability to respond to MTWs and small-scale contingencies (SSCs). The predominant planning scenarios for MTWs are a conflict in Northeast Asia (NEA) involving North and South Korea, and conflict in SWA involving Iraq. Concatenation of these scenarios leads to the two MTW criteria for force sizing. That is, a stressing case for U.S. planning involving two nearly simultaneous major theater wars. Other major theater conflicts can be imagined, such as a resurgent Russia or a militarily aggressive China, as well as conflicts involving India, Iran, etc. From the standpoint of the scorecard, each of these is represented by a group of columns in which the subordinate effectiveness measures involve variations in scenarios or cases within scenarios, and, subordinate to those, measures of conflict outcome.[1]

The U.S. ability to deal with SSCs provides another set of criteria. Again, one has scenarios such as Bosnia, Panama, Haiti, and Somalia, variations in conditions associated with each (with or without allies, various constraints on the use of force, etc.), and measures of outcome such as friendly casualties and time to subdue hostile elements. In examining stressing cases for force structure it is possible to consider one or more of the SSCs happening simultaneously with an MTW.

A hierarchy of columns (effectiveness measures) for the Respond component (capability for contingencies) might then look as follows:

1. Top Level (strategy component)

   Respond (to contingencies)

2. Level 2 (scenarios)

   NEA

   SWA

   NEA & SWA

   NEA & Bosnia

   Bosnia

   Haiti

   Etc.

3. Level 3 (cases of scenarios)

   (under SWA)

[1]The cases chosen for the scorecard reflected extensive scenario-space exploration such as described in Davis, Hillestad, and Crawford (1997).

Late reaction by U.S.

Restricted access to Persian Gulf

Long buildup

Etc.

4. Level 4 (measures of scenarios)

(under each scenario case)

Time to stop initial attack

Distance penetrated by enemy

Friendly casualties

Time to restore border

Time to defeat enemy and destroy warmaking potential

Etc.

A single DynaRank scorecard provides the capability to represent up to three of these levels of criteria. However, by linking subordinate scorecards it is possible to create as many levels of criteria as desired. For example, in the QDR work we used two levels of scorecard, with the first level representing all components of the QDR strategy and the subordinate scorecards representing cases of scenarios, measures of outcome, and other subordinate measures. We did not in fact address SSCs in our work.

The Shaping Component of Strategy

Environment shaping involves activities such as maintaining presence through carrier operations or in-place forces, development of alliances, joint training with allies, prepositioning equipment, basing agreements, and diplomacy. In addition, previous use of force that demonstrates the willingness to react and the maintenance of modern ready forces also shapes the strategic environment.[2] We broke down shaping activities by region, type of activity, and measure of that activity. We did not include diplomatic actions that did not involve defense activities or structure.

---

[2]This indicates that the strategy dimensions of respond, shape, and prepare now are not completely independent, so it is difficult to estimate how much of an option's costs should be allocated to the three components. For example, increasing forward naval presence contributes to both contingency capability and environment shaping. We credit both contributions, but we do not cost them separately. Some readers familiar with multi-attribute utility theory might argue that the objectives (strategy components) should be made independent so that separate costing would be possible. Others might argue that it is unfair to credit an option for its contributions in two categories, because that represents in some sense a "double counting." Upon reflection, we decided to proceed as we have—avoiding one set of problems by not separately costing by strategy component, and deliberately choosing to "double-count or triple-count" when an option has value in more than one strategy component.

Under these assumptions, the hierarchy for shaping is as follows:

1. Top Level (strategy component)

   Shape the security environment

2. Level 2 (region)

   Europe

   Central/South America

   Africa

   Mideast

   Northeast Asia

   Etc.

3. Level 3 (environment shaping activities)

   Naval presence

   Stationed forces

   Basing agreements

   Prepositioned equipment

   Joint exercises

   Etc.

4. Level 4 (measures of shaping activities)

   (under Naval presence)

   Carrier days in region

   Port visits by Naval forces

   Other Naval presence (noncarrier activities)

   (under stationed forces)

   Quantity of U.S. ground forces deployed

   Quantity of U.S. air forces deployed

   Etc.

The "Prepare Now" Component of Strategy

This dimension of strategy is concerned with actions to prepare the U.S. military for the future, in particular actions that go beyond preparing contingency capability for scenarios that are being addressed more routinely under the “respond” component, which is largely oriented toward the midterm and currently identifiable threats. We have called this “strategic adaptiveness” and have considered the various aspects of the future that require us to undertake activities now. Force modernization is one aspect of this. As opponents develop new capabilities, such as modern precision weapons, accurate long-range missiles, and weapons of mass destruction, it is essential that U.S. forces be prepared to cope with them. First, modernization might be measured against futuristic scenarios, or more subjectively judged by degree of ability to cope with enemy technology advances or technology breakthroughs—against threats that go beyond what we can already see for the midterm. A second aspect of the future is the changing strategic environment. If base access is reduced or alliances disappear, it may be necessary for U.S. forces to operate from longer distances, perhaps even conducting operations from the continental United States (CONUS) or space. A third aspect that must be considered is the potential effect of alternative future defense budgets. It is possible, in the absence of strong actions by Congress, that entitlement spending will cause dramatically reduced defense budgets and that this will in turn cause large changes in force structure or, alternatively, if force structure is protected, this will curtail the ability of the forces to modernize. Certain actions taken now, such as reengineering Army brigades, Naval battle groups, and Air Force squadrons, might enable services to better withstand severe budget shocks. The criteria for the prepare now component of strategy are as follows:

1. Top Level (strategy component)

   Prepare now (strategic adaptiveness)

2. Level 2 (component)

   Technological adaptiveness

   Strategic adaptiveness

   Budget adaptiveness

3. Level 3 (component)

   (Under technological adaptiveness)

   Information warfare

   Threats to space activities

   Weapons of mass destruction

   Etc.

   (Under strategic adaptiveness)

   Greatly reduced access to airbases and ports

   Chinese dominance in NEA

Etc.

(Budgetary adaptiveness)

Reducible infrastructure

Graceful reductions in force structure

Etc.

Typically, for our initial work we used subjective judgments about how the various modernization, force structure, and efficiency options would affect each of these criteria. It is possible that a more extensive effort could develop specific scenarios or cases and attempt to model or otherwise measure the extent to which certain options affect these criteria.

DEFINING AND CATEGORIZING THE OPTIONS

We considered as options a wide variety of programs calling for acquisition, reengineering, or reducing forces (or defense agencies). They included options to reduce Army, Air Force, and Navy force structure in various ways; options to acquire new capabilities such as advanced aircraft and tactical missile defense; and options to reengineer forces for additional efficiency or effectiveness or both. The options we evaluated were drawn from (but did not include all the ideas of) future vision documents of the DoD (Shalikashvili, 1996; Barnett, 1996; Defense Science Board, 1995, 1996; DDR&E, 1996), the Army (Army TRADOC, 1996); the Navy (CNO, 1997; and National Research Council, 1997), and the Air Force (O'Hanlon, 1995; MacGreggor, 1997), as well as informal discussions with experts and independent papers by outside analysts.[3] We organized the options by whether their primary focus was force structure, modernization, or efficiency (which includes reengineering options). Figure 3.1 shows some of the elements of this categorization. By and large, the options were ones that seemed at least plausible by 2010, the nominal last year for our assessments. Informally, we have considered the 2020 era, but we do not discuss that analysis here.

Why did we choose this level of representation of the options? There are multiple reasons. First, many items involved the force elements being discussed during the QDR (e.g., carrier battle groups). Second, we chose some items based on the availability of cost and effectiveness data at this level of resolution and on the fact that many of these options were considered in previous analysis at RAND and elsewhere. Lastly, additional options were chosen to provide a reasonably broad coverage of the choices available to the services and DoD. However, we do not claim that the options were an exhaustive representation of those choices. We examined 60 options across the above categories.

---

[3]We were also acquainted generally with ideas later published in the *Strategic Assessment 1997* by National Defense University's Institute for National Strategic Studies. Because of our midterm focus, however (out to perhaps 2010), we did not in fact represent some of the more radical restructurings well. Also we did not represent some of the ideas from other sources mentioned here that focus on the longer run and possible military peer competitors.

RANDMR996-3.1

Force Structure	Modernization	Efficiencies
Option	Option	Option
Reduce by x Army divisions	Reduce F-22 buy by .25x (subst.)	Increase squadron size for AF
Reduce by x AF wings	Local TMD	Reengineered brigades
Reduce by x CVBGs	Boost Phase TMD	Privatize and reengineer DoD medical system
Reduce by x Trident subs	Smart-ship tech. for x CVBGs	Sortie rate improvements
...	Replace x CVBGs by CCBGs	Mediterranean home porting for x CVBGs
	...	

TMD = tactical missile defenses.
CCBGs = contingency-capable battle groups.

Figure 3.1—Options Were Organized in Force Structure, Modernization, and Efficiency Categories

COSTS, BUDGETS, AND BASELINES

For most of our analysis, we used annualized costs of the individual options as the sole measure of cost. That is, we used only one cost dimension and that was obtained by amortizing acquisition costs over the expected life of the system, adding operating and support costs, and discounting future costs. This is admittedly crude, and certainly not enough resolution for programming choices, but was considered enough to allow us to obtain first-order estimates of cost for cost-effectiveness estimates. As a baseline, we used the 2001 Program Objective Memorandum (POM) forces and budget assumptions.

COST AND EFFECTIVENESS EVALUATION OF THE OPTIONS

We evaluated the options with a mix of quantitative and subjective models, and subjective judgments.[4] Campaign models provided the MTW outcomes, games and

[4]A subjective model explicitly identifies key considerations and distinguishes among cases. A "subjective assessment" might be something like assuming—without explanation—that a much smaller force structure, even if armed with high-technology precision weapons, would have serious problems in a protracted land war in Asia. Another example is assigning a high subjective value for a theater missile defense system because of the weapons of mass destruction (WMD) threat. We did not, in this particular study, actually use models to compute the consequences. In prior analyses and games, however, that subjective value was manifest.

expert judgment provided effectiveness estimates for capabilities in SSCs, simple presence models provided measures of presence for the shaping component, and subjective judgments were used to fill in the remainder of the scorecard. We used a hierarchy of scorecards to make the assessments so that contingency capabilities (MTW, SSCs, and multiple MTWs and SSCs) were represented in one scorecard, shaping activities were represented in another, and the aggregate results of these assessments were rolled up into a strategy scorecard and combined with subjective assessments of strategic adaptiveness in the main scorecard. Figures 3.2 through 3.4 show portions of the scorecards that were used for several elements of the hierarchy. Although the analysis was merely a prototype, and although we would now do a more in-depth analysis, the results were instructive and in some cases seemed credible.

Parts of the scorecards not shown include the cost dimensions (annualized costs), the remaining options, and the additional regions for the shaping scorecard. Each scorecard had an identical list of 60 options of the types described above.

WEIGHTING MEASURES AND RANKING THE OPTIONS

Utilizing these scorecards, we next examined the effects of weighting different components of the criteria in the different scorecards on the cost-effectiveness ranking of options. We attempted to determine whether there were robust options, ranking high regardless of the weighting, and whether there were options strongly dependent on weighting-specific scenarios, measures, or strategy components. We also used the approximate cumulative graphs to determine the accumulation of effectiveness and cost/savings as we selected the highest ranked options in order of ranks. The following are the general conclusions we reached by this process:

- The options in the efficiency category appear to be the most robust to changes in emphasis on criteria.
- The best modernization options address "Achilles' Heels" or make hard objectives feasible.
- The justification for modest reductions in active force structure is fairly robust where such options are politically feasible.
- Important strategic options involve reengineering.

ROBUST CAPABILITIES

We looked at the effect of varying the emphasis or the weighting of criteria on the cost-effective ranking of options. For example, the highest ranked options fell into the efficiency category. Figure 3.5 shows that with high weight on the shaping component of strategy and with equal weights on all components of strategy, most of the same options appear at the top of the list when ranked by cost-effectiveness. These types of comparisons can be shown using the DynaRank Rank display.

RANDMR996-3.2

Top Level QDR Scorecard

	Top Level Measures		Capability for Contingen-cies (1)							Environ-ment Shaping	Strategic Adaptive-ness (1)		Wt	Annual-ized Cost/ Savings (1)	Total Cost
	Mid Level Measures		Europe (1)	SWA (1)		E. Asia (1)		2 MRCs (1)			Tech	Other	Wt		
	Base Level Measures		Poland (1)	Late (1)	Long (1)	DPRK (1)	China (1)	Std. (1)	Late (1)				Wt		
	Goal -->		100	100	100	100	100	100	100	100	100	100	100		
Option No.	Option	Flag	100	59	80	100	100	100	20	71	80	60	74.5	0	0
32	Double weapon effective-ness of 20 wings	1	100	64	98	100	100	100	80	73	90	70	82	4.4	4.4
48	12 Reengineer-ed brigades	1	100	60.3	92	100	100	100	60	72	88	68	79.8	-1.2	-1.2
49	13 Reengineer-ed brigades	1	100	60	92	100	100	100	60	72	88	68	79.7	-1.3	-1.3
28	C4ISR enhance-ments	1	100		90	100	100	100	40		90	70		0.4	0.4
27	TBMD	1	100		80	100	100	100	30		90	70			0.75
31	Double weapon effective-ness of 10 wings	1	100	64	94	100	100	100	60	73	84	64	79	2.2	2.2
20	AF prepo (2 theaters)	1	100	61	90	100	100	100	90	73	80	60	78.7	1	1

Wt

Figure 3.2—A Portion of the Main QDR Scorecard

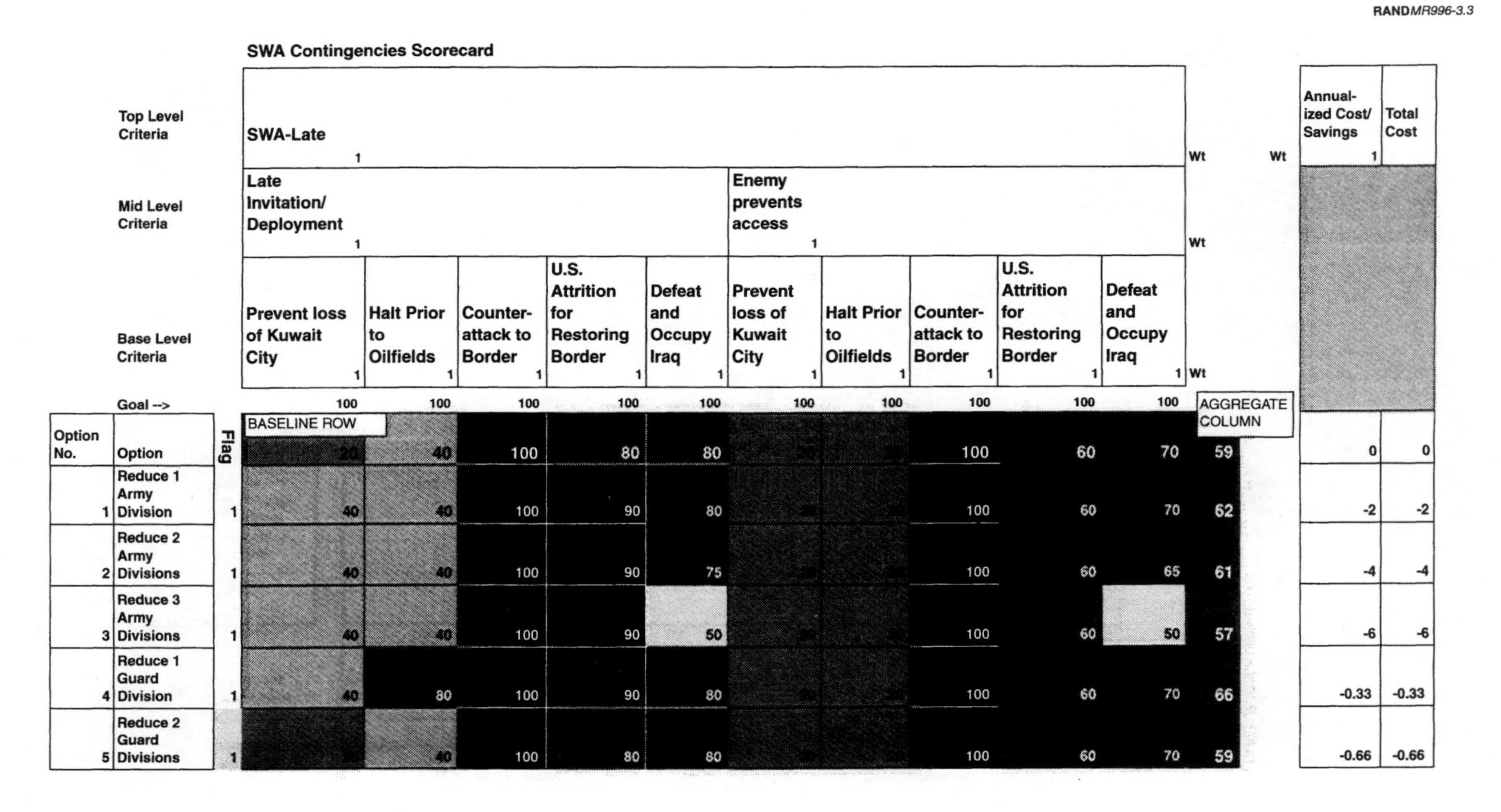

RANDMR996-3.3

SWA Contingencies Scorecard

			Late Invitation/Deployment (Wt 1)					Enemy prevents access (Wt 1)							
Top Level Criteria			SWA-Late (Wt 1)												
Base Level Criteria			Prevent loss of Kuwait City (1)	Halt Prior to Oilfields (1)	Counter-attack to Border (1)	U.S. Attrition for Restoring Border (1)	Defeat and Occupy Iraq (1)	Prevent loss of Kuwait City (1)	Halt Prior to Oilfields (1)	Counter-attack to Border (1)	U.S. Attrition for Restoring Border (1)	Defeat and Occupy Iraq (1)	AGGREGATE COLUMN	Annual-ized Cost/ Savings (1)	Total Cost
Goal -->			100	100	100	100	100	100	100	100	100	100			
Option No.	Option	Flag	BASELINE ROW 20	40	100	80	80	20	[illegible]	100	60	70	59	0	0
1	Reduce 1 Army Division	1	40	40	100	90	80	[illegible]	[illegible]	100	60	70	62	-2	-2
2	Reduce 2 Army Divisions	1	40	40	100	90	75	[illegible]	[illegible]	100	60	65	61	-4	-4
3	Reduce 3 Army Divisions	1	40	40	100	90	50	[illegible]	[illegible]	100	60	50	57	-6	-6
4	Reduce 1 Guard Division	1	40	80	100	90	80	[illegible]	[illegible]	100	60	70	66	-0.33	-0.33
5	Reduce 2 Guard Divisions	1	[illegible]	40	100	80	80	[illegible]	[illegible]	100	60	70	59	-0.66	-0.66

Figure 3.3—A Portion of the Subordinate Scorecard Used for the SWA Contingencies (See Plate IV for color illustration)

RANDMR996-3.4

Shaping Scorecard

			Europe 1										Greater Middle East 1		
	Mid Level Measures		Overseas Presence 1				Power Projection Capabilities 1			Security Alliances and Coalitions 1			Overseas Presence 1		
	Base Level Measures		Deployed Forces 1	Infrastructure incl. Prepositioning 1	Security Assistance Inc. FMI 1	OOTW 1	Rapidly Deployable Forces 1	Deployable within Months 1	Sustainability 1	Forces 1	Operational Prowess 1	Cohesiveness and Planning 1	Deployed Forces 1	Infrastructure incl. Prepositioning 1	Security Assistance Inc. FMI 1
	Goal -->		100	100	100	100	100	100	100	100	100	100	100	100	100
Option No.	Option	Flag													
	BASELINE ROW		100	100	100	80	60	100	100	100	80	80	60	80	60
1	Reduce 1 Army Division	1	90	100	100	80	60	100	100	100	80	80	60	80	60
2	Reduce 2 Army Divisions	1	82	100	100	80	60	100	100	100	80	80	56	80	60
3	Reduce 3 Army Divisions	1	82	100	100	80	60	100	100	100	80	80	52	80	60
4	Reduce 1 Guard Division	1	100	100	100	80	60	100	100	100	80	80	60	80	60

Figure 3.4—A Portion of the QDR Scorecard Used for the Shaping Component of Strategy (See Plate V for color illustration)

RANDMR996-3.5

Robustness of Efficiency Options (Shaded Options)

Option Rank—High Shaping Wt.	Option Rank—Equal Wts.
• Replace 3 CVBG with 3 CCBG	• Replace 3 CVBG with 3 CCBG
• Med home port for CVBG	• 12 Reengineered brigades
• Smart ship technology	• Med home port for CVBG
• 12 Reengineered brigades	• Reduce 3 Guard Divisions
• Reduce 3 Guard Divisions	• DOD Agency reduction
• DOD Agency reduction	• Larger AF Squadrons
• Larger AF Squadrons	• Smart ship technology
• ATBM	• .75 F22 Buy
• ---	• Reduce ships in 3 CVBG
	• ---

Figure 3.5—Comparison of Ranked Lists with Different Emphasis on the Strategy Components

Some options represented cost savings with essentially no effect on performance, while others improved performance and saved money. For example, the availability of a Mediterranean home port would not only improve the ability to maintain naval presence, but would also save money in terms of the number of carriers needed to provide the increased presence.[5] A postulated reengineering of the Army to move to a corps/brigade structure with modern, smaller brigades and a smaller overall support structure could improve the ability to deploy quickly and mount a more substantial early defense at lower cost as a result of smaller manpower requirements. There might or might not be penalties for environment shaping, and there would be penalties in large manpower-intensive wars. Higher combat sortie rates for aircraft permit the same force to be more lethal and permit the job to be done with fewer deployed aircraft. These types of capabilities, while often showing small marginal improvements, are important because of their "win-win" nature and the fact that cumulatively, they can add up to large improvements in capability at a very reasonable cost. Figure 3.6 shows a DynaRank cumulative plot of option costs and effectiveness with the options added one at time from left to right and with the highest ranked options by cost-effectiveness added first.

[5]Obviously, adding a homeport would not have the same benefit as an additional battle group for fighting two long MTWs in which the United States lacked adequate land bases. However, we did not highlight such a case in our effectiveness measures because it seems relatively less important than the cases we used or than the ability to maintain peacetime presence. That, of course, is a judgment.

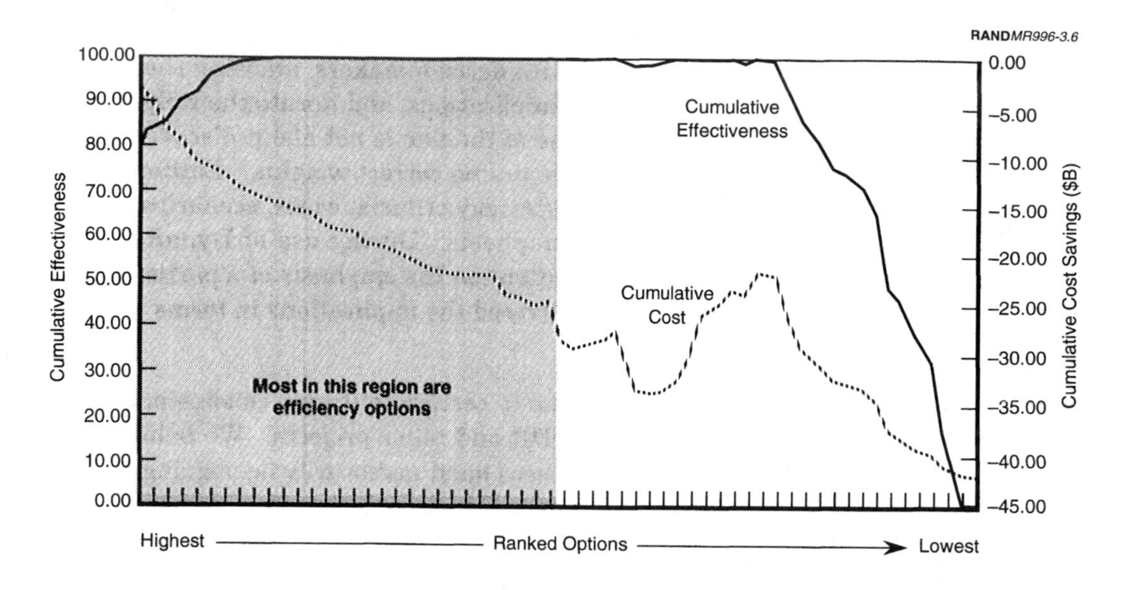

Figure 3.6—Efficiency Options Save Money and Often Improve Effectiveness as Well

Note that the options that improve effectiveness and save money are chosen first. Effectiveness builds up and savings accrue at the same time. Next, options that save money and do not reduce effectiveness are accumulated. Effectiveness stays constant while savings continue to accrue. It turned out that many of the options in this category are also efficiency options.

OBSERVATIONS ABOUT THE PROTOTYPE APPLICATION OF DYNARANK

DynaRank is just one of many possible tools for resource-allocation analysis of force structure and force capability. It is, however, the only tool we know of that can integrate the analysis of the contributions of capabilities to all components of the new defense strategy. And it is one of few tools that permits the direct integration of cost and effectiveness in the evaluation of many disparate policy options. It requires that the analyst be explicit about objectives and criteria and, in turn, it permits a dynamic, interactive investigation of the choices, allowing variation of emphasis on the criteria at any stage in the hierarchy of measurement from individual scenario outcome measure, to emphasis on scenario, to emphasis on strategy component. It permits the linkage of a detailed representation of an option to operations and ultimately to the high-level defense strategy. As such, it can be an information tool as well as an analysis tool, capturing what is known about linkages, performance, and costs.

The primary purpose of the tool is to assist the discovery of rationale about policy options and classes of options. Standard use includes looking for robust capabilities, attempting to identify the cases in which certain options are important, trying to screen out some capabilities, and mixing performance with cost. We believe that DynaRank should be used in a dialogue with decisionmakers, allowing them to select the emphasis on criteria, observe the implications, and iterate the weighting to develop the rationale for a choice. The use of the tool is not about discovering the "correct weights" because we believe there are no correct weights. Rather, by examining the consequences of weights on strategy criteria, cases, scenarios, etc., it is possible to study the implications of emphasis. During use of DynaRank, we have observed decisionmakers who were intent on the emphasis of a particular criteria change that emphasis once they discovered the implications in terms of policy choices.

We have illustrated how DynaRank led us to certain tentative conclusions about policy choices during the period of the QDR and other projects. We believe that this is not a one-time analysis and the process must necessarily be ongoing as budgets change, new capabilities become apparent, or new threats, scenarios, and cases become important. *Thus, we also believe that embedding this type of tool and process within DoD, perhaps as part of the PPBS process, is important to maintain the rationalization for the new defense strategy, protecting transformation process, and shaping activities as threats change, budgets decrease, or people attempt to protect force structure at the expense of modernization.*

Finally, we note that our analysis was merely a prototype. In some cases we postulated both new forces and their effectiveness; in other cases, cost estimates were quite soft. Further, we did not really do justice to the issue of preparing for the longer term (e.g., 2020) or to the subtleties of environment shaping and strategic adaptiveness. Nonetheless, some of the conclusions seem robust and others at least point toward what should be checked in more depth. And, ultimately, it was a prototype effort without the benefit of close interactions with policymakers. In future experimental work, we intend to increase substantially interactions with policymakers, and with DoD and service staffs.

Appendix A

DYNARANK TUTORIAL

CREATING A SCORECARD

This appendix goes through the steps of creating and using a scorecard within DynaRank. We have also embedded most of these instructions in a "README" worksheet within DynaRank that annotates the functions and provides some step-by-step instructions.

The Dimensions of a Scorecard

The primary dimensions of a scorecard are the policy options or alternatives to be evaluated, the measures used to evaluate those options, the dimensions of cost of the alternatives, and the evaluation or body of the scorecard. Virtually everything else in the scorecard is derived from these dimensions. Prior to describing how to build a DynaRank scorecard and applying the scorecard tools, we give some specific examples of options, measures, and cost dimensions.

The Options and Classes of Options

Options are whatever the user chooses to evaluate. They may represent individual actions and objects under consideration, or they may include packages of more-detailed options. The scorecard permits a second level of definition to support categorizing the options. This level is called the option type. For example, Table A.1 shows a list of defense planning options categorized into modernization, efficiency, and force structure options.

The option type is often useful in forming general conclusions about the types of options that rank highest when the DynaRank ranking processes are used.

Table A.2 shows defense options considered as packages and Table A.3 shows options considered in a transportation study.

The Hierarchies of Measures

The measures are used to evaluate the options. In most cases the measures exist in hierarchies—from general to more specific. For example, the hierarchy in

Table A.1

Options Grouped by Option Type for a Scorecard

Option Type	Option
Modernization	Arsenal ship
Modernization	ATBM system
Modernization	Equip F-22s for anti-armor missions
Modernization	C4ISR package
Modernization	Rapid SEAD
Modernization	Standoff munitions
Efficiency	Allied defense package (helos/ATACMS/advisors)
Efficiency	Faster tacair deployment
Efficiency	Double-surge sortie rates
Force Structure	Minus 2 TFWs
Force Structure	Minus 1 active division
Force Structure	Minus 1 CVBG
Modernization	25% fewer F-22s
Efficiency	New BRAC
Efficiency	Med. home port
Efficiency	Forward-deployed wing
Combined	Arsenal ship plus allied package

Table A.2

Defense Package Options and Types

Option Type	Option
Efficiency	High-sortie-rate air units
Efficiency	Reengineered battle groups
Efficiency	Enhanced mobility package
Modernization	C4I package
Modernization	Weapons modernization
Infrastructure	Infrastructure cuts
Infrastructure	Ground force structure cuts

Table A.3

Transportation Planning Options and Types

Option Type	Option
Efficiency	Reward truck driver for fuel savings
Efficiency	Telecommunications
Efficiency	Intermodal facilities
Direct Mitigation	Special truck lanes
Direct Mitigation	Cleaner diesel engines
Infrastructure	Freight rail line
Infrastructure	New highway sections

Table A.4 begins with the three components of defense strategy at the top, has scenarios or subordinate measures of the strategy component at the next level, and specific cases at the next lower level.

Table A.4

Measures Hierarchy for Defense Strategy

Portfolio Component	Level 2 Component	Level 3 Component
Contingency capability (Respond)	SWA	Case 1
		Case 2
	2 MTWs or MTW & MOOTW	
Environment shaping (Shape)	Overseas presence	
	Security assistance	
Strategic adaptiveness (Prepare now)	Transforming the force	
	Other hedges	

It is frequently desirable to continue the hierarchy down additional steps. In the case in Table A.4, a second scorecard might be used. For example, the SWA scenario might be described by the two cases listed and for each case we may be interested in multiple measures of the cases as shown in Table A.5.

Similarly, an analyst might be interested in measuring options in the hierarchy of measures for a transportation problem shown in Table A.6.

Table A.5

SWA Measures Hierarchy

Top-Level Measures	Mid-Level Measures	Base-Level Measures
SWA	Case 1—Surprise attack	Enemy penetration at halt
		Defeating and occupying Iraq
	Case 2—Delayed deployment	Enemy penetration at halt
		Defeating and occupying Iraq

Table A.6

Measures Hierarchy for a Transportation Problem

Top-Level Measures	Mid-Level Measures	Base-Level Measures
Emissions	CO_2	
	C_xH_x	HydroCarbon 1
		HydroCarbon 2
	SO^2	
	Particulates	
Safety	Fatalities	Drivers
		Passengers
	Accidents	Trucks
		Autos
	Haz. goods	
Congestion		
Economics	Added value	
	Employment	Transportation
		Other
Noise		

The Costs and Dimensions of Costs

The cost dimensions are used for various purposes in the DynaRank scorecards. For example, an analyst could represent the costs of options during various periods of time. Weighting of these costs could be used to identify how various options are sensitive to the period of time when they are paid for. Or, weights can be used to discount the costs in future time periods. Table A.7 shows dimensions of costs by period.

Alternatively, cost components can be represented as those in Table A.8. A third possibility is to include components that represent the cost uncertainty. In this case, weights on these components could be lighter.

Using a Template to Create a Scorecard

DynaRank builds a new scorecard from a template. The first step is to obtain a blank template to use in defining the dimensions of the new scorecard. The DynaRank menu item **Template** and the submenu item **New** are selected by the user from the scorecard tools menu, **SC Tools**. Figure A.1 illustrates this selection from among the other Excel menu options. If the scorecard to be built is similar to another, it may be more efficient to modify an existing template. In this case the **Duplicate** option of the submenu is used, and DynaRank will create a copy of the existing template for user modification. The **Clear** option removes the information from a template. Templates can be removed from the worksheet by using the Excel submenu item, **Delete Sheet**, in the **Edit** menu.

After **New** is selected, a dialogue box will appear and request that the template be given a name. The default name, New Template, can be used if desired. Figure A.2 shows the dialogue box with a template name entered. It is useful for keeping track of templates and scorecards to append a T or Template to the template name.

Table A.7

Measures of Cost—
By Time Period

Cost Dimension
Cost (1–10 years)
Cost (11–20 years)
Cost (21–30 years)

Table A.8

Measures of Cost—Cost
Components for New Systems

Cost Dimension
Research and development
Acquisition
Operations and support

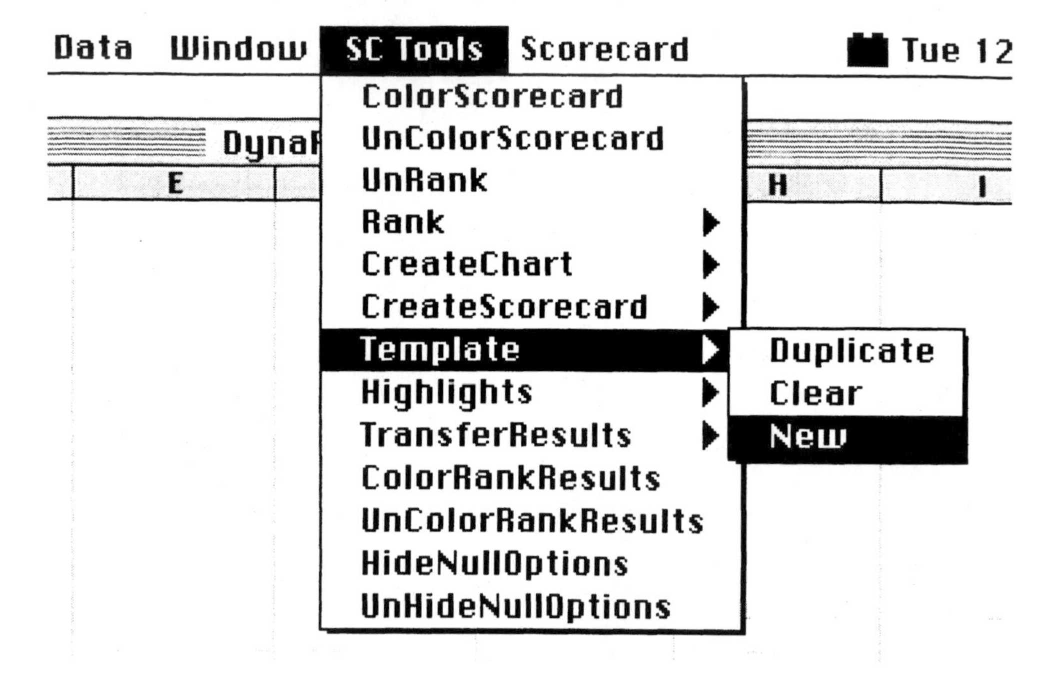

Figure A.1—Using the Pull-Down Menu to Create a New Scorecard Template

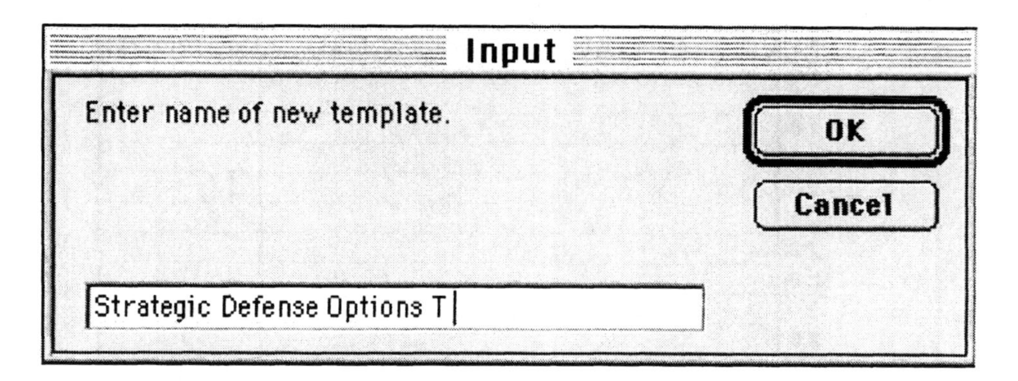

Figure A.2—Dialogue Box for Naming a Scorecard Template

After the user provides the name and selects OK in the dialogue box, DynaRank will create a blank template with three major blocks for entering the options and types, the measures, and the cost dimensions, respectively. It will also include a block for labels or titles to be placed on the scorecard. Figure A.3 shows the option and option type blocks along with the labels block. Note that the options are automatically numbered. These numbers are not accessible to the user, but the options can be entered in any desired order. The option type block is used to define the type or category of the option and is often useful in highlighting the contributions of options of specific types. The labels blocks are used only to insert titles on the scorecard and can be left blank. They do not necessarily correspond to the name of the scorecard worksheet (which we will get to in a moment).

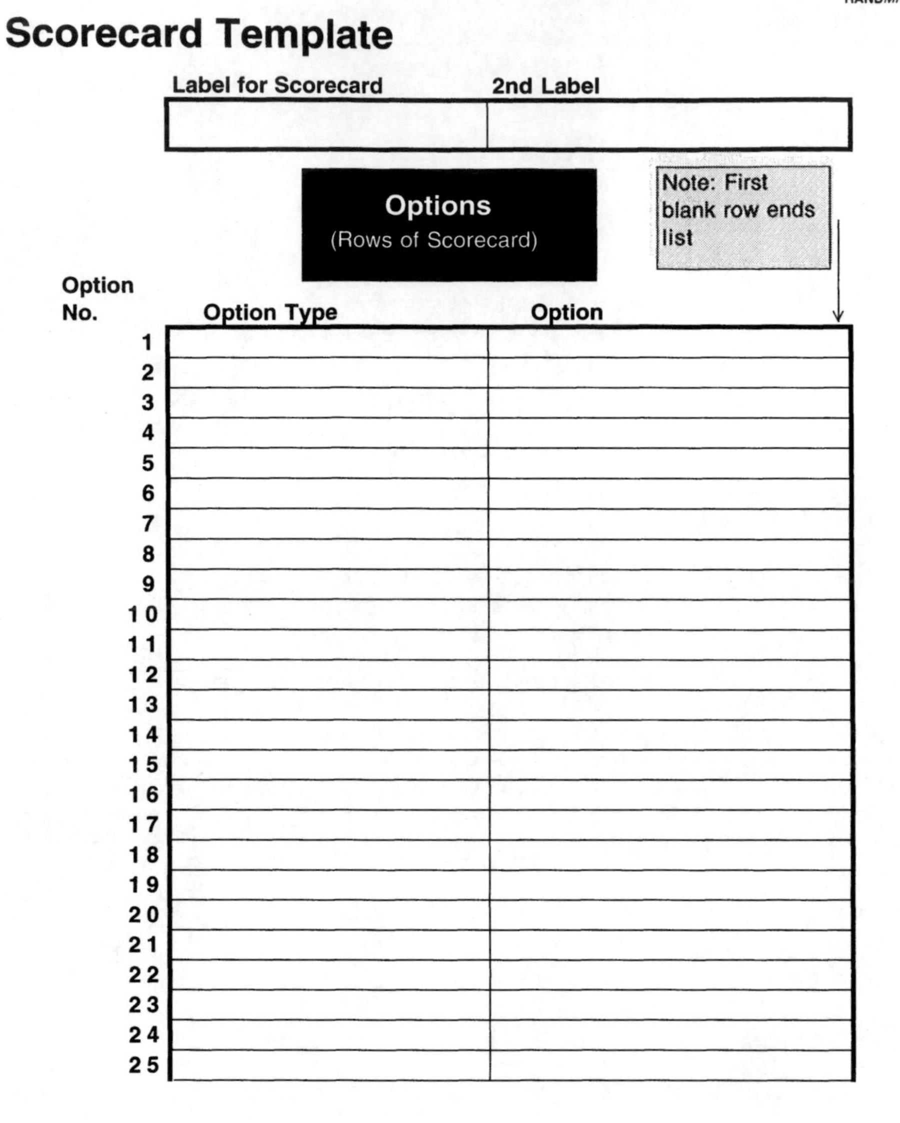

Figure A.3—DynaRank Template Blocks for Entering Options, Option Types, and Scorecard Labels

Figure A.4 illustrates a filled-in option and option type template block.

The measures blocks are assumed to be hierarchical, and the hierarchy can be three levels deep. Figure A.5 illustrates a measures block. In the template these appear as rows, but when the scorecard is created this list will form the columns of the scorecard. The titles of the measures levels in the first row can be left at the default or changed by the user if desired.

Figure A.6 shows this block filled in for a specific scorecard. Note that in contrast to the option rows, the measures are entered with blanks and order does makes a

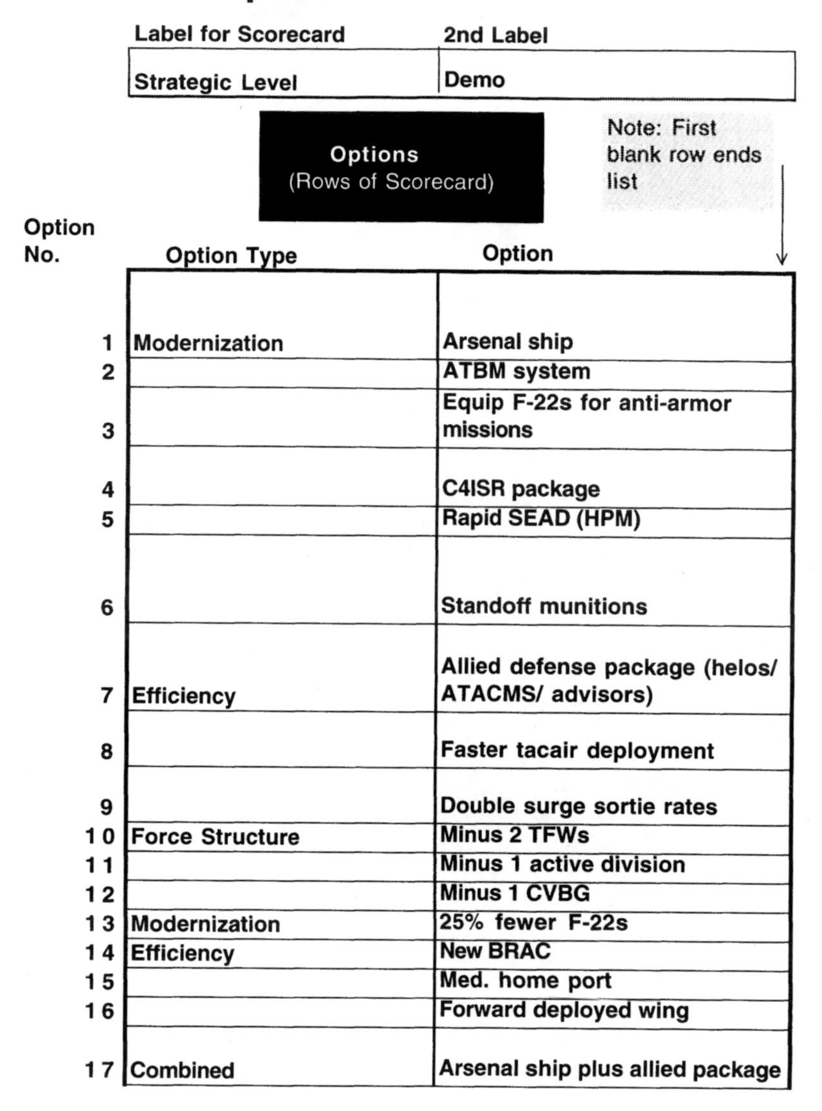

RANDMR996-A 4

Scorecard Template

Label for Scorecard	2nd Label
Strategic Level	Demo

Options (Rows of Scorecard)

Note: First blank row ends list

Option No.	Option Type	Option
1	Modernization	Arsenal ship
2		ATBM system
3		Equip F-22s for anti-armor missions
4		C4ISR package
5		Rapid SEAD (HPM)
6		Standoff munitions
7	Efficiency	Allied defense package (helos/ ATACMS/ advisors)
8		Faster tacair deployment
9		Double surge sortie rates
10	Force Structure	Minus 2 TFWs
11		Minus 1 active division
12		Minus 1 CVBG
13	Modernization	25% fewer F-22s
14	Efficiency	New BRAC
15		Med. home port
16		Forward deployed wing
17	Combined	Arsenal ship plus allied package

Figure A.4—Filled-In Option and Option Type Blocks in a DynaRank Template

difference. In fact, it is important that the user enter the measures in this manner for the DynaRank scorecard to operate properly. First, a top-level measure is entered. Then, if there is a level 2 measure associated with this top-level item, it is entered on the same row. Similarly, if there is a level 3 measure for the level 2 measure, it is also entered on the same row. The next row would show only another level 3 measure associated with the first level 2 measure, etc. The key is to carry the measures down as far as they go in the hierarchy on the same row, do not enter a level 2 or 3 measure more than once for the same next highest

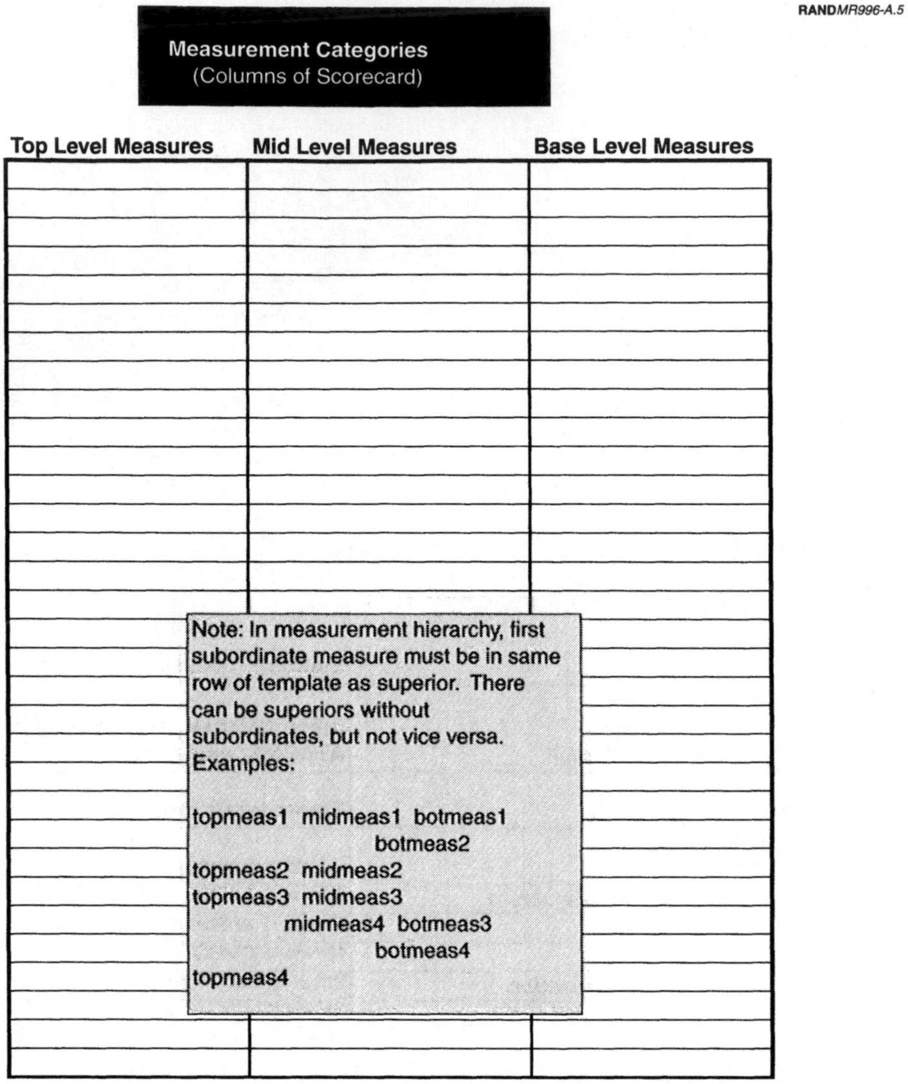

RANDMR996-A.5

Measurement Categories
(Columns of Scorecard)

Top Level Measures Mid Level Measures Base Level Measures

Note: In measurement hierarchy, first subordinate measure must be in same row of template as superior. There can be superiors without subordinates, but not vice versa.
Examples:

topmeas1 midmeas1 botmeas1
botmeas2
topmeas2 midmeas2
topmeas3 midmeas3
midmeas4 botmeas3
botmeas4
topmeas4

Figure A.5—The Measures Block in a DynaRank Template

measure. Also, do not leave blank rows in the measures block until all measures have been entered. DynaRank assumes that the first blank row indicates the end of the measures. The measures will appear in the scorecard with the first level 1 measure in the first measures column of the scorecard and in the topmost row.

The last block of the template corresponds to the cost dimensions desired in the scorecard. These also will become columns in the scorecard. Figure A.7 illustrates a filled-in cost block. These cost components will appear with the top component in the leftmost column of the scorecard.

Color Blank Cells

High Color Value: 100

Low Color Value: 0

CE Cost Wt.: 1

Strategic Level Scorecard RAND*MR996-P1*

	Contingency Capability (RESPOND) 1			Environment Shaping (SHAPE) 1		Strategic Adaptiveness (PREPARE NOW) 1		Wt or Min	Costs: Annualized Cost 1
Top Level Measures ->								Wt	Wt
Mid Level Measures ->	SWA 1		2 MTWs or MTW& MOOTW 1	Overseas Presence 1	Security Assistance 1	Transforming the Force 1	Other Hedges 1	Wt	
Base Level Measures ->	Case 1 1	Case 2 1						Wt	
Goal -->	100	100	100	100	100	100	100		

Graph			Option	Flag	Case 1	Case 2	2 MTWs or MTW& MOOTW	Overseas Presence	Security Assistance	Transforming the Force	Other Hedges	Aggregate Column	Annualized Cost	Total Cost
			Base case row		65.9	40.0	75.0	65.0	65.0	65.0	65.0	64.7	0	0
1	1	Modernization	Arsenal ship	1	71.3	40.0	79.0	67.0	65.0	70.4	65.0	67.0	0.251	0.251
1	2	Modernization	ATBM system	1	65.9	[illegible]	85.0	60.0	65.0	65.0	65.0	69.0	1	1
1	3	Modernization	F-22s anti-armor	1	67.1	40.0	82.0	65.0	65.0	66.2	65.0	66.1	0.05	0.05
1	4	Modernization	C4ISR package	1	64.9	[illegible]	82.0	65.0	65.0	80.0	65.0	70.1	0.4	0.4
1	5	Modernization	Rapid SEAD (HPM)	1	79.2	40.0	85.0	65.0	65.0	74.3	65.0	68.7	0.11	0.11
1	6	Modernization	Standoff munitions	1	73.7	40.0	80.0	65.0	65.0	72.8	65.0	67.4	0.2	0.2
1	7	Efficiency	Allied pkg	1	96.2	84.2	95.0	75.0	75.0	80.0	65.0	80.0	0.3	0.3
1	8	Efficiency	Faster tacair deployment	1	68.0	40.0	79.0	65.0	65.0	67.2	65.0	65.9	0.105	0.105
1	9	Efficiency	Double surge sortie rates	1	72.4	40.0	82.0	65.0	65.0	71.5	65.0	67.5	0.05	0.05
1	10	Force Structure	Minus 2 TFWs	1	65.9	40.0	62.0	61.0	65.0	65.0	65.0	61.8	-0.7	-0.7

Plate I—A DynaRank Scorecard (See Figure 2.5 on p.18 and discussion)

RAND*MR996-P2*

View A	View B	View C
Emphasizes Contingencies	Emphasizes Shaping	Emphasizes Cost
New BRAC	**25% fewer F-22s**	**Minus 1 active division**
Double surge sortie rates	**New BRAC**	**Minus 1 CVBG**
Allied defense package (helos/ATACMS/ advisors)	**Allied defense package (helos/ATACMS/ advisors)**	**25% fewer F-22s**
Equip F-22s for anti-armor missions	**Forward deployed wing**	**Minus 2 TFWs**
Rapid SEAD (HPM)	**Med. home port**	**New BRAC**
Forward deployed wing	**Arsenal ship plus allied package**	**Forward deployed wing**

Plate II—Alternative "Views" Stored in a Rank Sheet with Common Items Colored Alike (See Figure 2.8 on p. 22 and discussion)

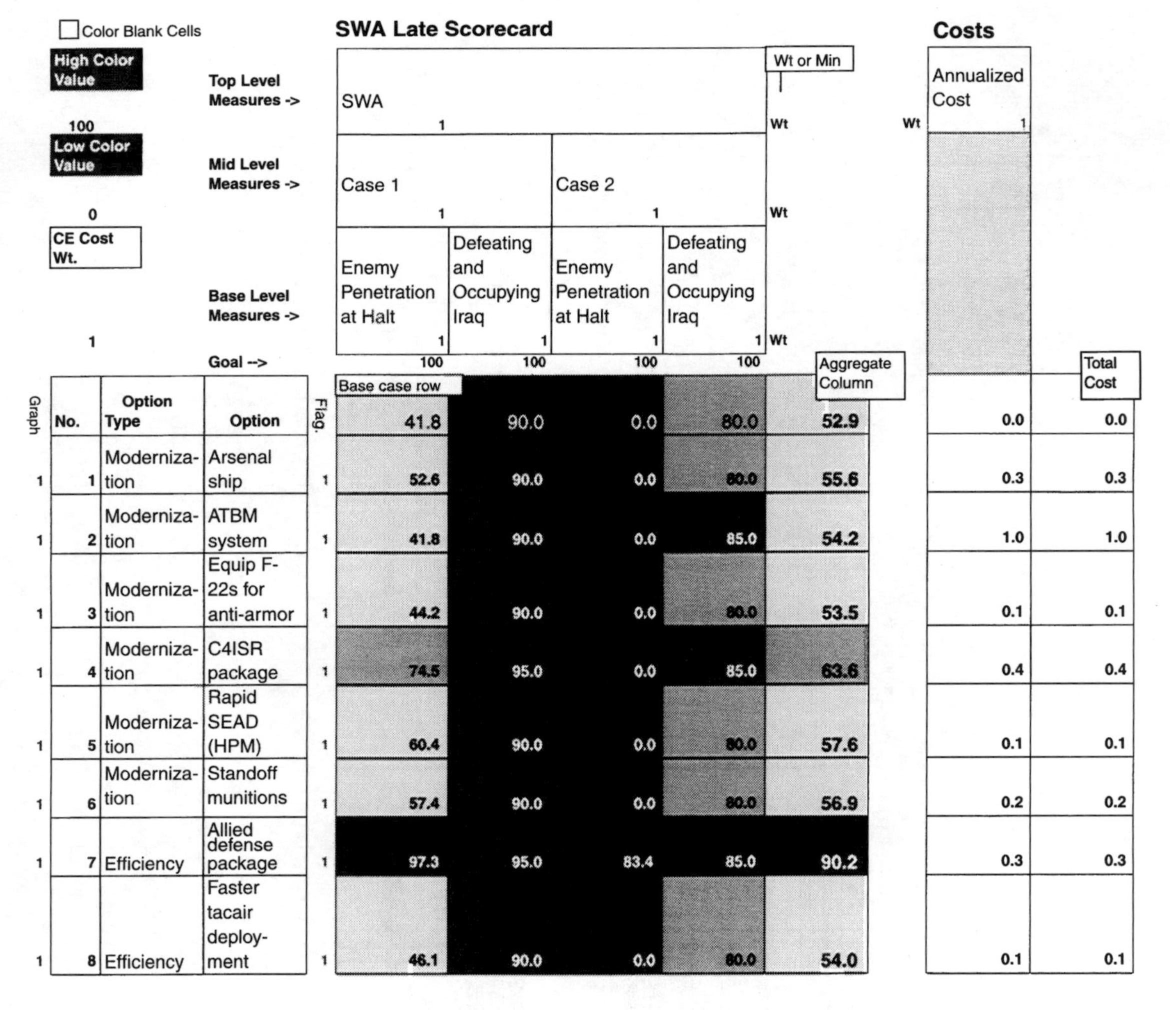

RAND*MR996-P3*

SWA Late Scorecard

Color Blank Cells

High Color Value: 100

Low Color Value: 0

CE Cost Wt.: 1

Top Level Measures -> SWA (1)

Mid Level Measures -> Case 1 (1); Case 2 (1)

Base Level Measures ->

Goal -> 100 / 100 / 100 / 100

Graph	No.	Option Type	Option	Flag.	Case 1: Enemy Penetration at Halt (1)	Case 1: Defeating and Occupying Iraq (1)	Case 2: Enemy Penetration at Halt (1)	Case 2: Defeating and Occupying Iraq (1)	Aggregate Column (Wt or Min)	Costs: Annualized Cost (1)	Total Cost
					Base case row 41.8	90.0	0.0	80.0	52.9	0.0	0.0
1	1	Modernization	Arsenal ship	1	52.6	90.0	0.0	80.0	55.6	0.3	0.3
1	2	Modernization	ATBM system	1	41.8	90.0	0.0	85.0	54.2	1.0	1.0
1	3	Modernization	Equip F-22s for anti-armor	1	44.2	90.0	0.0	80.0	53.5	0.1	0.1
1	4	Modernization	C4ISR package	1	74.5	95.0	0.0	85.0	63.6	0.4	0.4
1	5	Modernization	Rapid SEAD (HPM)	1	60.4	90.0	0.0	80.0	57.6	0.1	0.1
1	6	Modernization	Standoff munitions	1	57.4	90.0	0.0	80.0	56.9	0.2	0.2
1	7	Efficiency	Allied defense package	1	97.3	95.0	83.4	85.0	90.2	0.3	0.3
1	8	Efficiency	Faster tacair deployment	1	46.1	90.0	0.0	80.0	54.0	0.1	0.1

Plate III—A Subordinate Scorecard (See Figure 2.9 on p. 24 and discussion)

RAND*MR996-P4*

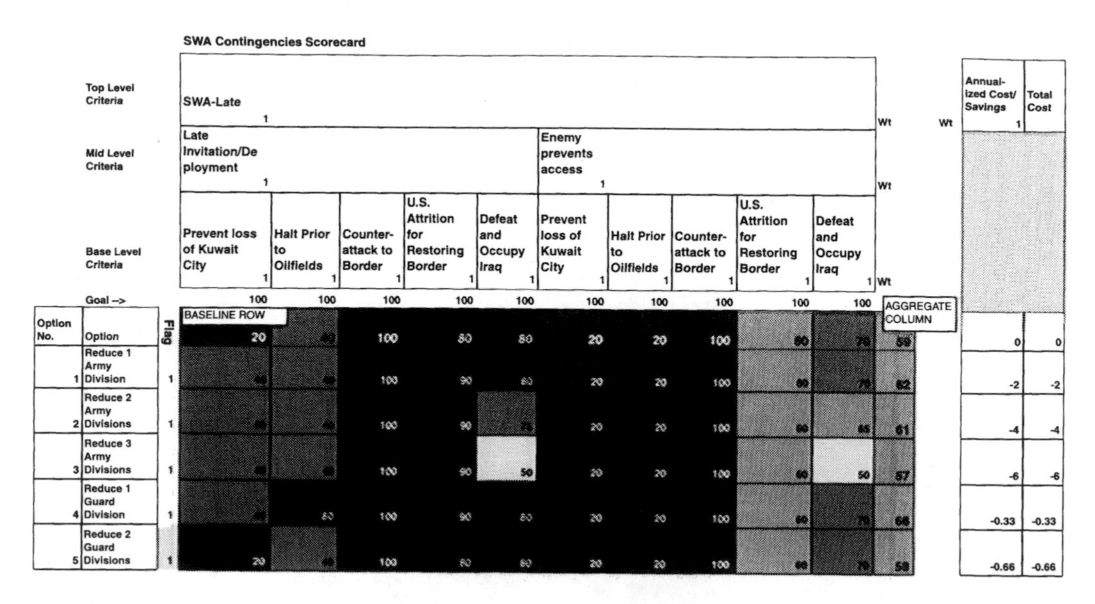

SWA Contingencies Scorecard

			SWA-Late (1)												
Top Level Criteria			SWA-Late 1										Wt	Wt	Annual-ized Cost/ Savings 1	Total Cost
Mid Level Criteria			Late Invitation/Deployment 1					Enemy prevents access 1					Wt			
Base Level Criteria			Prevent loss of Kuwait City 1	Halt Prior to Oilfields 1	Counter-attack to Border 1	U.S. Attrition for Restoring Border 1	Defeat and Occupy Iraq 1	Prevent loss of Kuwait City 1	Halt Prior to Oilfields 1	Counter-attack to Border 1	U.S. Attrition for Restoring Border 1	Defeat and Occupy Iraq 1	Wt			
Goal -->			100	100	100	100	100	100	100	100	100	100	AGGREGATE COLUMN			
Option No.	**Option**	**Flag**	BASELINE ROW													
			20	40	100	80	80	20	20	100	60	70	59		0	0
1	Reduce 1 Army Division	1	40	40	100	90	80	20	20	100	60	70	62		-2	-2
2	Reduce 2 Army Divisions	1	40	40	100	90	75	20	20	100	60	65	61		-4	-4
3	Reduce 3 Army Divisions	1	40	40	100	90	50	20	20	100	60	50	57		-6	-6
4	Reduce 1 Guard Division	1	40	80	100	90	80	20	20	100	60	70	66		-0.33	-0.33
5	Reduce 2 Guard Divisions	1	20	40	100	80	80	20	20	100	60	70	58		-0.66	-0.66

Plate IV—A Portion of the Subordinate Scorecard
(See Figure 3.3 on p. 34 and discussion)

RAND*MR996-P5*

Shaping Scorecard

Top Level Measures			Europe 1										Greater Middle East 1		
Mid Level Measures			Overseas Presence 1				Power Projection Capabilities 1			Security Alliances and Coalitions 1			Overseas Presence 1		
Base Level Measures			Deployed Forces 1	Infrastructure incl. Prepositioning 1	Security Assistance Inc. FMI 1	OOTW 1	Rapidly Deployable Forces 1	Deployable within Months 1	Sustainability 1	Forces 1	Operational Prowess 1	Cohesiveness and Planning 1	Deployed Forces 1	Infrastructure incl. Prepositioning 1	Security Assistance Inc. FMI 1
Goal -->			100	100	100	100	100	100	100	100	100	100	100	100	100
Option No.	**Option**	**Flag**	BASELINE ROW												
			100	100	100	80	60	100	100	100	80	80	60	80	60
1	Reduce 1 Army Division	1	90	100	100	80	60	100	100	100	80	80	60	80	60
2	Reduce 2 Army Divisions	1	82	100	100	80	60	100	100	100	80	80	56	80	60
3	Reduce 3 Army Divisions	1	82	100	100	80	60	100	100	100	80	80	52	80	60
4	Reduce 1 Guard Division	1	100	100	100	80	60	100	100	100	80	80	60	80	60

Plate V— A Portion of the QDR Scorecard Used for the Shaping Component of Strategy
(See Figure 3.4 on p. 35 and discussion)

RAND*MR996-P6*

Strategic Level Scorecard

☐ Color Blank Cells

High Color Value: 100

Low Color Value: 0

CE Cost Wt.: 1

					Contingency Capability (RESPOND) 1			Environment Shaping (SHAPE) 1		Strategic Adaptiveness (PREPARE NOW) 1		Wt or Min: Wt	Costs	
			Top Level Measures ->		Contingency Capability (RESPOND) 1			Environment Shaping (SHAPE) 1		Strategic Adaptiveness (PREPARE NOW) 1		Wt	Annualized Cost (Wt 1)	
			Mid Level Measures ->		SWA 1		2 MTWs or MTW & MOOTW 1	Overseas Presence 1	Security Assistance 1	Transforming the Force 1	Other Hedges 1	Wt		
			Base Level Measures ->		Case 1 1	Case 2 1						Wt		
			Goal -->		100	100	100	100	100	100	100	Aggregate Column		Total Cost
Graph			Option	Flag.	65.9 (Base case row)	40.0	75.0	65.0	65.0	65.0	65.0	64.7	0	0
1	1	Modernization	Arsenal ship	1	71.3	40.0	79.0	67.0	65.0	70.4	65.0	67.0	0.251	0.251
1	2	Modernization	ATBM system	1	65.9	42.5	85.0	80.0	65.0	65.0	65.0	69.0	1	1
1	3	Modernization	F-22s anti-armor	1	67.1	40.0	82.0	65.0	65.0	66.2	65.0	66.1	0.05	0.05
1	4	Modernization	C4ISR package	1	84.8	42.5	82.0	65.0	65.0	80.0	65.0	70.1	0.4	0.4
1	5	Modernization	Rapid SEAD (HPM)	1	75.2	40.0	85.0	65.0	65.0	74.3	65.0	68.7	0.11	0.11
1	6	Modernization	Standoff munitions	1	73.7	40.0	80.0	65.0	65.0	72.8	65.0	67.4	0.2	0.2
1	7	Efficiency	Allied pkg	1	98.2	84.2	95.0	75.0	75.0	80.0	65.0	80.0	0.3	0.3
1	8	Efficiency	Faster tacair deployment	1	68.0	40.0	79.0	65.0	65.0	67.2	65.0	65.9	0.105	0.105
1	9	Efficiency	Double surge sortie rates	1	72.4	40.0	82.0	65.0	65.0	71.5	65.0	67.5	0.05	0.05
1	10	Force Structure	Minus 2 TFWs	1	65.9	40.0	62.0	61.0	65.0	65.0	65.0	61.8	-0.7	-0.7

Plate VI—A DynaRank Scorecard (See Figure A.11 on p. 51 and discussion)

RAND*MR996-P7*

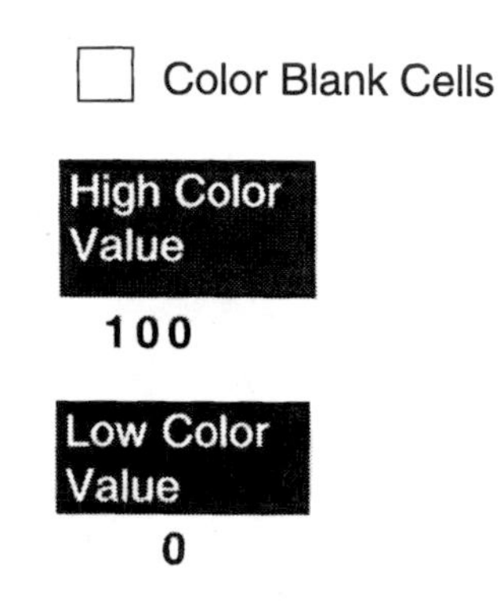

Plate VII—Color Values (See Figure A.18 on p.59 and discussion)

RAND*MR996-P8*

Scorecard-Aggregation to Top Level Measures

Lvl1 Wts.	1	1	1
Option	Contingency Capability (RESPOND)	Environment Shaping (SHAPE)	Strategic Adaptiveness (PREPARE NOW)
BaseCase	64.0	65.0	65.0
Arsenal ship	66.0	66.0	67.7
ATBM system	69.0	72.5	65.0
Equip F-22s for anti-armor missions	67.5	65.0	65.6
C4ISR package	67.5	65.0	72.5
Rapid SEAD (HPM)	69.0	65.0	69.7
Standoff munitions	68.1	65.0	68.9
Allied defense package (helos/ ATACMS/ advisors)	74.3	75.0	72.5

Plate VIII—Aggregate Scorecard for Top-Level Measures (See Figure A.22 on p.64 and discussion)

RANDMR996-P9

Scorecard-Aggregation to Mid Level Measures

Lvl2 Wts.	1	1	1	1	1	1
	Contingency Capability (RESPOND)		Environment Shaping (SHAPE)		Strategic Adaptiveness (PREPARE NOW)	
Option	SWA	2 MTWs or MTW & MOOTW	Overseas Presence	Security Assistance	Transforming the Force	Other Hedges
BaseCase	52.9	75.0	65.0	65.0	65.0	65.0
Arsenal ship	52.9	79.0	67.0	65.0	70.4	65.0
ATBM system	52.9	85.0	80.0	65.0	65.0	65.0
Equip F-22s for anti-armor missions	52.9	82.0	65.0	65.0	66.2	65.0
C4ISR package	52.9	82.0	65.0	65.0	80.0	65.0
Rapid SEAD (HPM)	52.9	85.0	65.0	65.0	74.3	65.0
Standoff munitions	56.2	80.0	65.0	65.0	72.8	65.0
Allied defense package (helos/ ATACMS/ advisors)	53.5	95.0	75.0	75.0	80.0	65.0

Plate IX—Derived Scorecard for Rollup to Level 2 Measures
(See Figure A.23 on p. 65 and discussion)

RANDMR996-P10

	Count of Ranks	Number for Top x		
	3	10		
24		**Contingency View**	**Shaping View**	**Adaptiveness View**
	1	**New BRAC**	**25% fewer F-22s**	**Minus 1 active division**
	2	**Double surge sortie rates**	**New BRAC**	**Minus 1 CVBG**
	3	**Allied defense package (helos/ ATACMS/advisors)**	**Allied defense package (helos/ ATACMS/advisors)**	**25% fewer F-22s**
	4	**Equip F-22s for anti-armor missions**	**Forward deployed wing**	**Minus 2 TFWs**
	5	**Rapid SEAD (HPM)**	**Med. home port**	**New BRAC**
	6	**Forward deployed wing**	**Arsenal ship plus allied package**	**Forward deployed wing**
	7	**Arsenal ship plus allied package**	**Arsenal ship plus allied package plus deployed wing**	**Double surge sortie rates**
		Arsenal ship plus allied package plus		

Plate X—Coloring the Common Items in the Top x Items of the Views
(See Figure A.32 on p. 73 and discussion)

RANDMR996-P11

RH Borders of Cells are Colors of Korea Risk SC

Color Blank Cells

		Contingency Capability (RESPOND) 1			Environment Shaping (SHAPE) 1	
Mid Level Measures ->		Korea 1		2 MTWs or MTW & MOOTW 1	Overseas Presence 1	Security Assistance 1
Base Level Measures ->		Case 1 1	Case 2 1			
Goal -->		100	100	100	100	100
Option	Flag.	Base case row 65.9	40.0	75.0	65.0	65.0
Arsenal ship	1	65.9	40.0	79.0	67.0	65.0
ATBM system	1	65.9	40.0	85.0	80.0	65.0
F-22s anti-armor	1	65.9	40.0	82.0	65.0	65.0
C4ISR package	1	65.9	40.0	82.0	65.0	65.0
Rapid SEAD (HPM)	1	65.9	40.0	85.0	65.0	65.0
Standoff munitions	1	72.4	40.0	80.0	65.0	65.0
Allied pkg	1	67.1	40.0	95.0	75.0	75.0
Faster tacair deployment	1	80.6	40.0	79.0	65.0	65.0
Double surge sortie rates	1	65.9	40.0	82.0	65.0	65.0
Minus 2 TFWs	1	68.0	40.0	62.0	61.0	65.0

(Row label for top level: **Top Level Measures ->**)

Plate XI—Highlights Shown on a Scorecard
(See Figure A.43 on p. 83 and discussion)

RANDMR996-A.6

Measurement Categories
(Columns of Scorecard)

Top Level Measures	Mid Level Measures	Base Level Measures
Contingency Capability (RESPOND)	SWA	Case 1
		Case 2
	2 MTWs or MTW&MOOTW	
Environment Shaping (SHAPE)	Overseas Presence	
	Security Assistance	
Strategic Adaptiveness (PREPARE NOW)	Transforming the Force	
	Other Hedges	

Figure A.6—A Filled-In Measures Block in a DynaRank Template

Creating a Scorecard from a Template

Once the template is completed, the user can cause the scorecard to be created by choosing the **New** submenu item of the **CreateScorecard** menu item in the **SC Tools** menu as illustrated in Figure A.8. The other submenu items of the CreateScorecard menu item will be discussed later. The New submenu item is only available from a scorecard template. (A template worksheet has an invisible column containing, among other data, the information that the worksheet is a scorecard template.)

At this point, the DynaRank system will request the name of the new scorecard or allow the use of the default. The dialogue box to name a scorecard is shown in Figure A.9. It is useful to append the word scorecard or SC to the end of the name to signify that the name belongs to a scorecard, to distinguish it from other types of worksheets that will appear in tabs at the bottom of the Excel workbook. It is possible to see names already in use for scorecards by pulling down the list in the

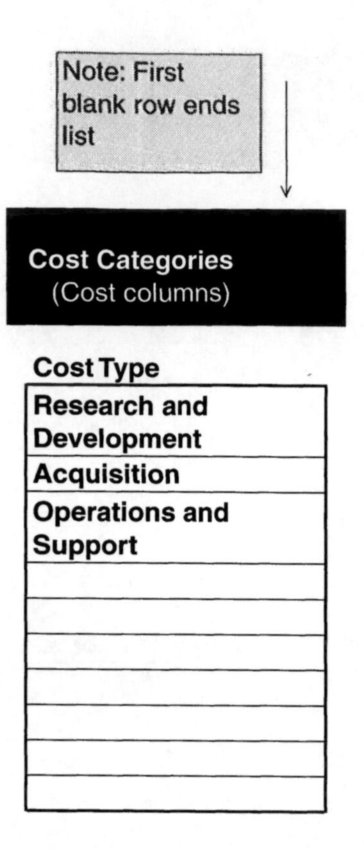

Figure A.7—A Filled-In Cost Block in a DynaRank Template

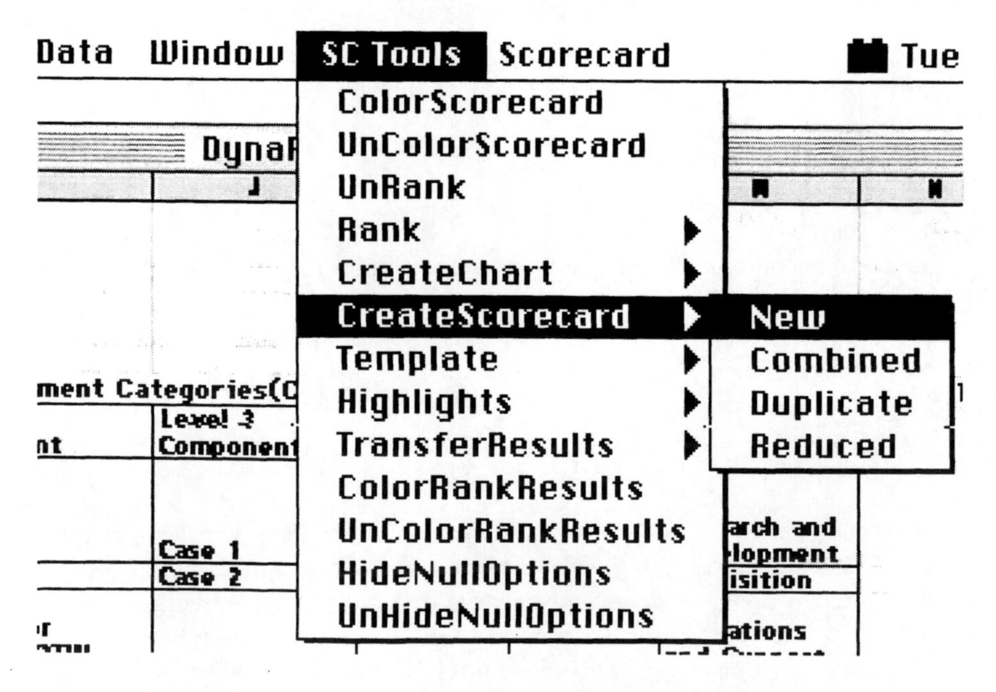

Figure A.8—Menu Selection from a Template Worksheet to Create a New Scorecard

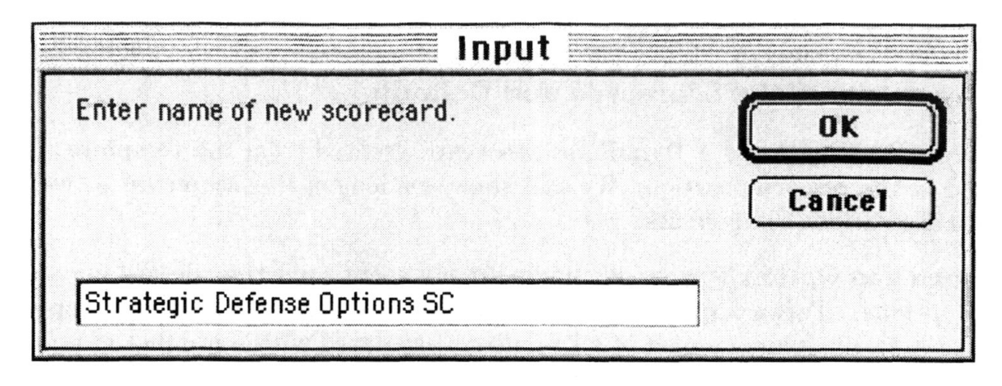

Figure A.9—Naming a New Scorecard

Scorecard menu on the Excel menu bar. This list is updated whenever scorecards are created or deleted and whenever the DynaRank model is opened. Selection of a scorecard in the menu of existing names will cause DynaRank to move to that scorecard worksheet. As with templates, scorecards and other special DynaRank worksheets have information stored to indicate what type of worksheet it is. An illustration of the pull down list of scorecard names appears in Figure A.10.

Once the name in the dialogue box has been entered and the user selects **OK**, DynaRank will create the desired scorecard with all of the functionality described in the next section. This process takes only a few minutes so that the user can quickly experiment with a number of variations of the scorecard as a problem is formulated, new options need to be added, criteria are changed, etc. An important aspect of and motivation for the DynaRank system was this flexibility in structuring and changing scorecards.

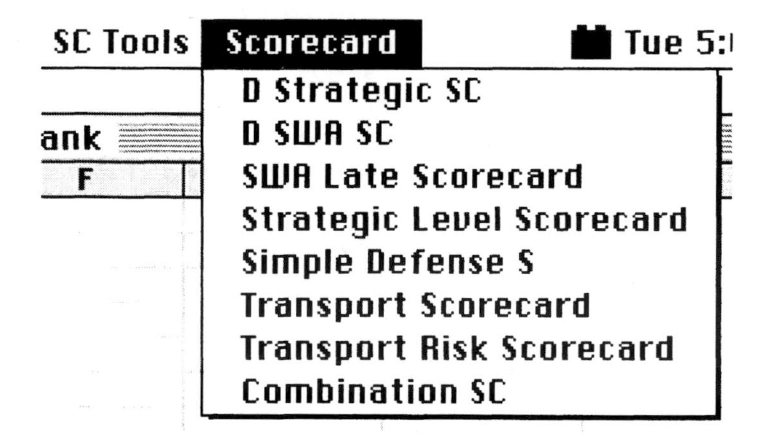

Figure A.10—Menu of Existing Scorecards in the Workbook

THE BASIC SCORECARD

Components of the Scorecard and Defaults

Figure A.11 illustrates a DynaRank scorecard created from the template illustrated in the previous section. We will show portions of this scorecard as we describe the various components.

Options and Option Types. We have already spent some time describing scorecard options. These appear as rows of the scorecard with one row per option. Figure A.12 illustrates a part of a DynaRank scorecard with a number of options as rows.

Option Inclusion Flag. The **Inclusion Flag**, shown in Figure A.12, permits a longer list of options to be included in the scorecard than might be used or evaluated for some purposes. For example, after some initial screening it may be desirable to prune the list of options for further consideration. This could be done by creating a new scorecard, but the DynaRank scorecard allows the Inclusion Flag to be set to 0 to exclude the option from consideration in most processing. If the user also elects to **Hide Null Options,** those with the Inclusion Flag set to 0 will have their rows hidden in the scorecard. Ranking processes will put these designated options at the bottom of a ranking. The default setting of the Inclusion Flag is 1, meaning that the option is to be considered. A value of 2 for the flag will force the options to the top of ranked lists regardless of their numerical evaluation. Some other values for the Inclusion Flag permit explicit groupings of options.

Measures and Weights. The measures define the columns of the basic DynaRank scorecard. Figure A.13 illustrates measures columns corresponding to the template shown earlier. The labels for the top-level measures appear on the first row, the default weights for these measures are the ones set immediately below in the next row, the level 2 measures labels make up the next row, followed by their default weights, etc. Note the structure of the measures columns as it relates to the template shown earlier. The first three columns are associated with the top-level measure, Contingency Capability (Respond). The first two columns are associated with the level 2 measure or scenario, SWA, under the Contingency Capability, and the first column is associated with case 1 of the SWA scenario. Note also how the measures structure is set up in the scorecard when the level 3 criteria are omitted. In this case, no measures appear at the level 3 row in the scorecard, and there are no weights for level 3. It is also possible to omit both level 2 and level 3 measures, including only the top-level measure.

Weights/Min Aggregation. A fundamental operation on the scorecard is to change the weights of the measures. This is done simply by changing the default weights to a new value as illustrated in Figure A.14. The weights are multiplied between levels so that, for example, if the weights were as shown in Figure A.14, each element in column 1 under Contingency Capability would be multiplied by the product of $0.25 \times 0.5 \times 0.5$. Each element of the last column, under Strategic Adaptiveness, would be multiplied by 0.25×0.5.

RANDMR996-A.11

☐ Color Blank Cells

High Color Value

100

Low Color Value

0

CE Cost Wt.

1

Strategic Level Scorecard

													Costs	
			Top Level Measures ->		Contingency Capability (RESPOND) 1			Environment Shaping (SHAPE) 1		Strategic Adaptive-ness (PREPARE NOW) 1		Wt or Min ↓ Wt	Wt Annualized Cost 1	
			Mid Level Measures ->		SWA 1		2 MTWs or MTW& MOOTW 1	Overseas Presence 1	Security Assistance 1	Transform-ing the Force 1	Other Hedges 1	Wt		
			Base Level Measures ->		Case 1 1	Case 2 1						Wt		
			Goal -->		100	100	100	100	100	100	100	Aggregate Column		Total Cost
Graph			**Option**	Flag.	Base case row 65.9	40.0	75.0	65.0	65.0	65.0	65.0	64.7	0	0
1	1	Moderniza-tion	Arsenal ship	1	71.3	40.0	79.0	67.0	65.0	70.4	65.0	67.0	0.251	0.251
1	2	Moderniza-tion	ATBM system	1	65.9	42.5	85.0	80.0	65.0	65.0	65.0	69.0	1	1
1	3	Moderniza-tion	F-22s anti-armor	1	67.1	40.0	82.0	65.0	65.0	66.2	65.0	66.1	0.05	0.05
1	4	Moderniza-tion	C4ISR package	1	84.8	42.5	82.0	65.0	65.0	80.0	65.0	70.1	0.4	0.4
1	5	Moderniza-tion	Rapid SEAD (HPM)	1	75.2	40.0	85.0	65.0	65.0	74.3	65.0	68.7	0.11	0.11
1	6	Moderniza-tion	Standoff munitions	1	73.7	40.0	80.0	65.0	65.0	72.8	65.0	67.4	0.2	0.2
1	7	Efficiency	Allied pkg	1	96.2	84.2	95.0	75.0	75.0	80.0	65.0	80.0	0.3	0.3
1	8	Efficiency	Faster tacair deployment	1	68.0	40.0	79.0	65.0	65.0	67.2	65.0	65.9	0.105	0.105
1	9	Efficiency	Double surge sortie rates	1	72.4	40.0	82.0	65.0	65.0	71.5	65.0	67.5	0.05	0.05
1	10	Force Structure	Minus 2 TFWs	1	65.9	40.0	62.0	61.0	65.0	65.0	65.0	61.8	-0.7	-0.7

Figure A.11—A DynaRank Scorecard
(See Plate VI for color illustration)

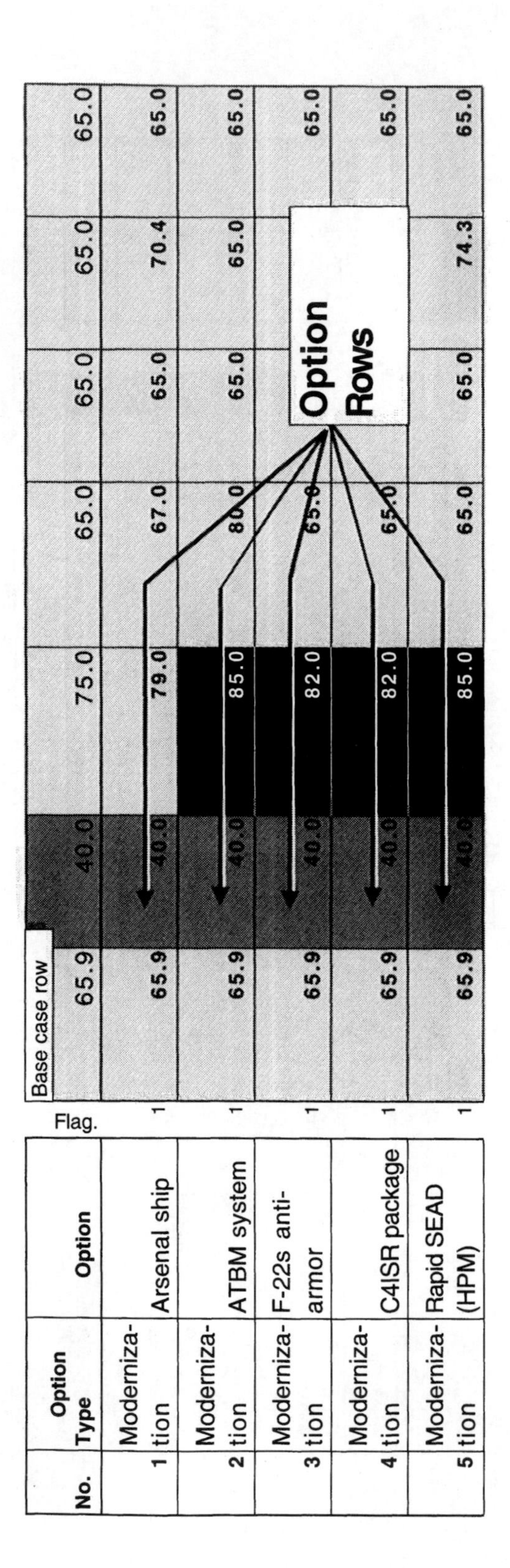

Figure A.12—Rows Are Associated with Options in the DynaRank Scorecard

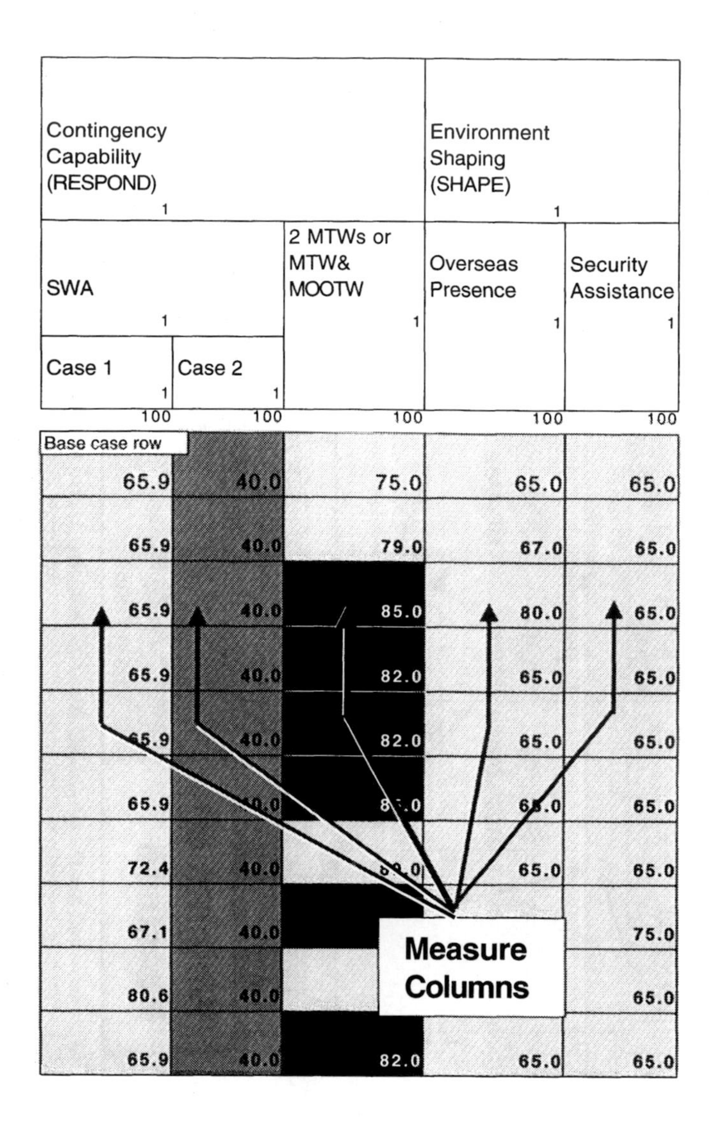

Figure A.13—Measures Form the Basic Columns of a DynaRank Scorecard

It is possible to use the worst case at any particular criteria level rather than taking the weighted average. The user can force this to happen by changing **Wt** in the Aggregate Utility Column to **Min,** as shown in Figure A.15. When this is done, all subcategories at the level designated with a Min will aggregate by picking the worst case from all cases within the next level criteria. For example, if the level 2 measures were designated with a Min in this scorecard, the worst case for Contingency Capability, the worst case for Environment Shaping, and the worst case for Strategic Adaptiveness would be combined with the weights at the top level.

Base-Case Row. There are two rows at the top of the scorecard, below the measures names and weights, that are not associated with options. The base-case row is used for values of outcomes when none of the options in the list are considered.

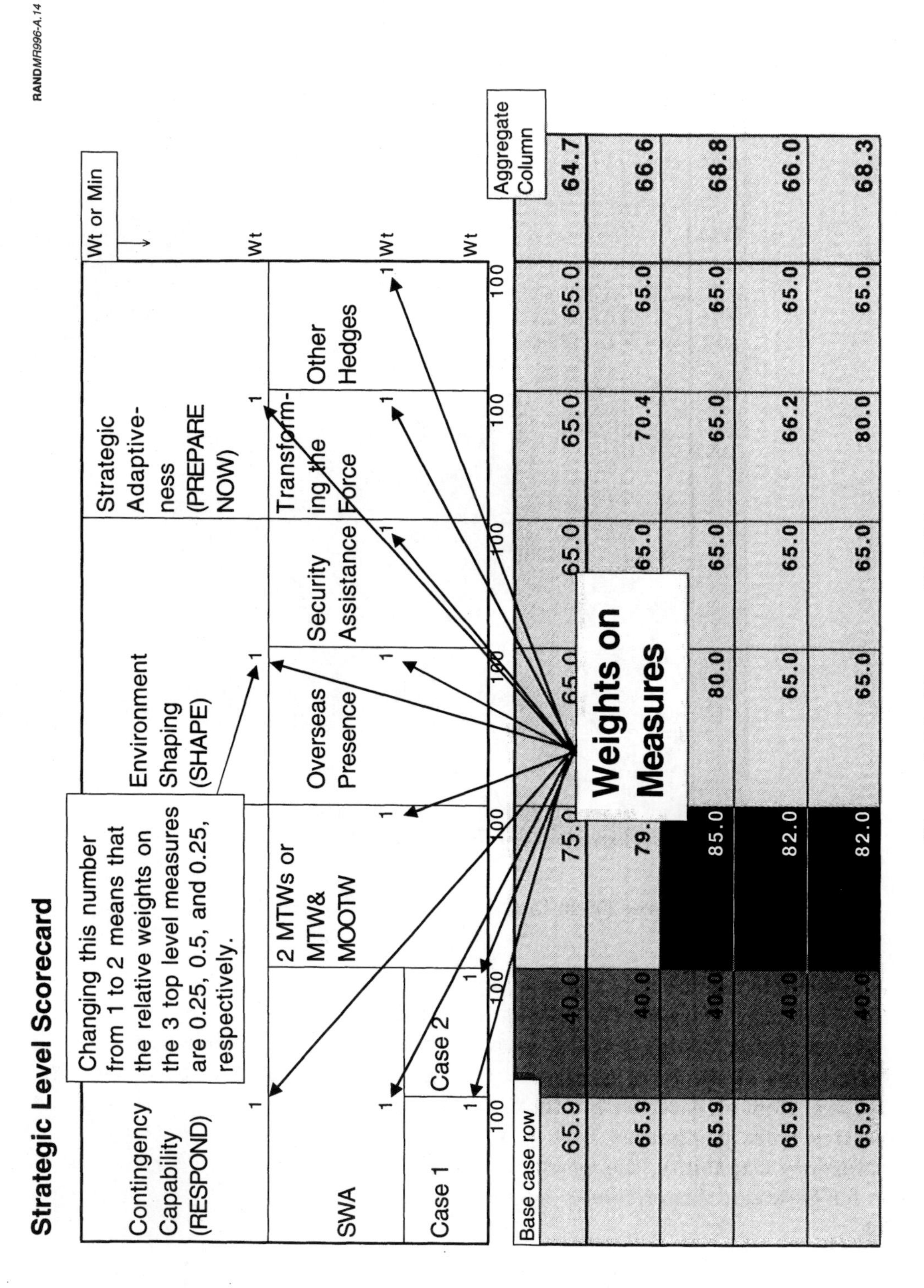

Figure A.14—Measures Weights

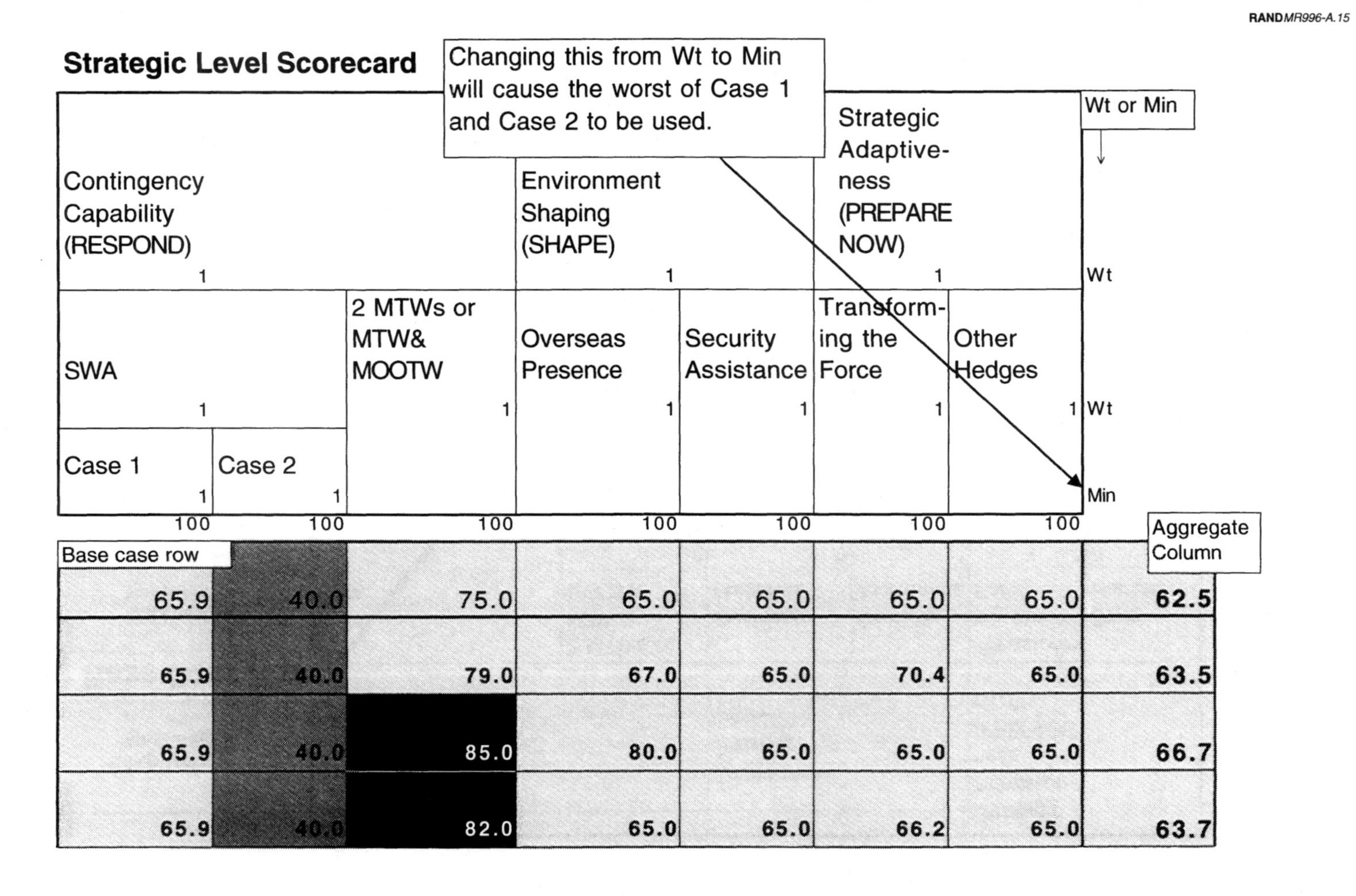

Contingency Capability (RESPOND) 1			Environment Shaping (SHAPE) 1		Strategic Adaptiveness (PREPARE NOW) 1		Wt	
SWA 1		2 MTWs or MTW& MOOTW 1	Overseas Presence 1	Security Assistance 1	Transform-ing the Force 1	Other Hedges 1	Wt	
Case 1 1	Case 2 1						Min	
100	100	100	100	100	100	100		
65.9	40.0	75.0	65.0	65.0	65.0	65.0		62.5
65.9	40.0	79.0	67.0	65.0	70.4	65.0		63.5
65.9	40.0	85.0	80.0	65.0	65.0	65.0		66.7
65.9	40.0	82.0	65.0	65.0	66.2	65.0		63.7

Figure A.15—Choosing the Min Across Measures

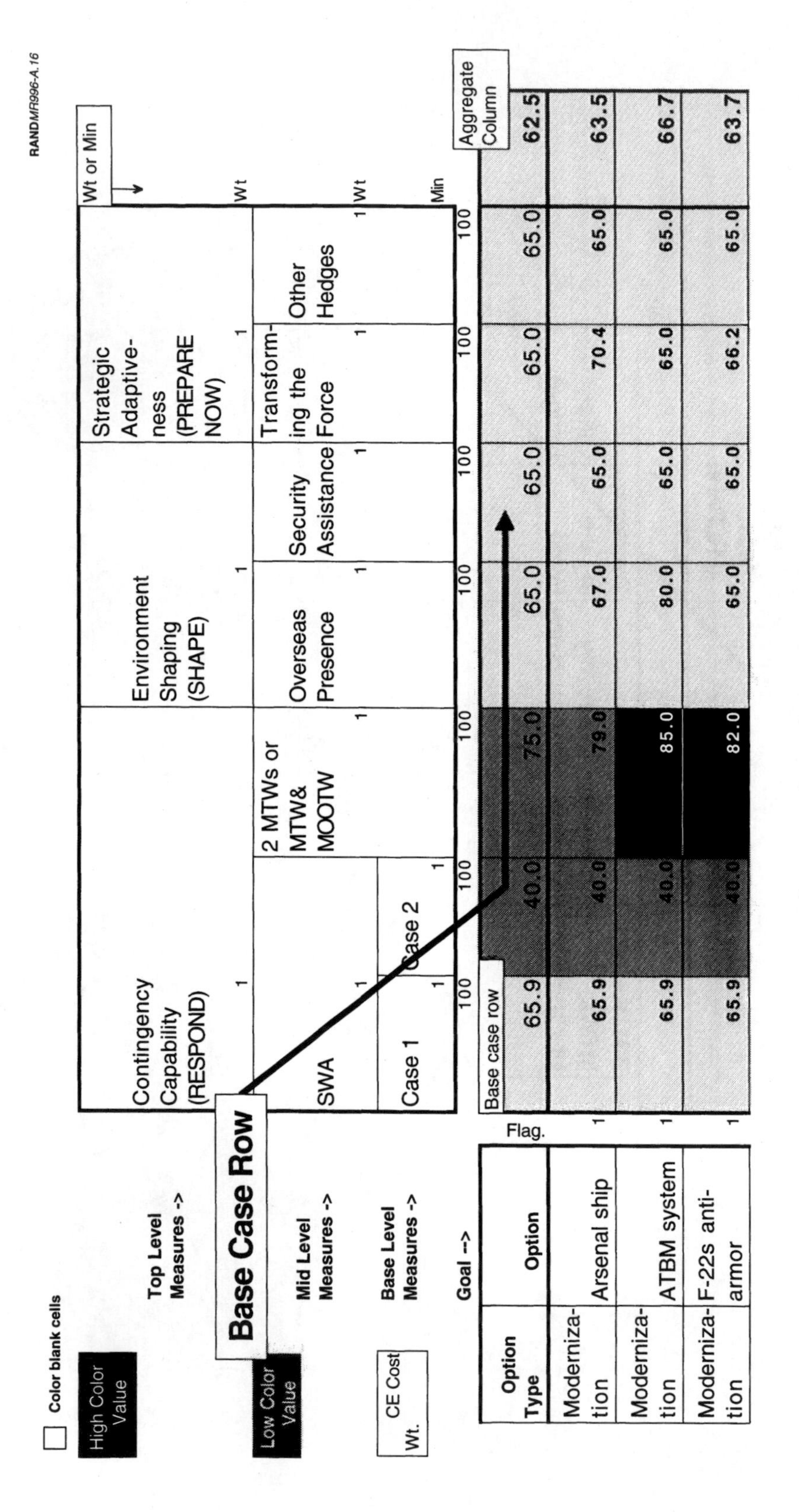

Figure A.16—The Base-Case Row of a Scorecard

It provides a reference point for how much is contributed by the option over and above the base case without the option. For accumulations, we assume that the difference between the row with an option added and this base-case row gives the marginal contribution of the option. Figure A.16 shows the base-case row for our scorecard example.

Goal Row. The second row at the top of the scorecard, immediately above the base-case row, is the goal row. The goals associated with each measure column are entered in this row and they define the desired performance for that measure. The default number is 100 (for 100 percent desired achievement in performance), but the goals can be set to any value desired for the problem. It is assumed in DynaRank that desired performance is at the high end of the numerical scale. Thus, the functions, models, or judgments that fill in the scorecard must perform a translation of actual outcomes to the scale used in the scorecard. This can generally be accomplished with simple outcome value or utility functions. Figure A.17 illustrates the goal row.

Color Values. For purposes of coloring the scorecard according to values of the options on measures, DynaRank needs to be told what number corresponds to the high color value (best value) and what number corresponds to the low color value (worst value). DynaRank automatically allocates bright red to a region around the low color value and bright green to a region around the high color value. The default values for high and low color values are 0 and 100, but the user may prefer a range of 1–10, 0–1, etc. There are three intermediate colors available, so the scale chosen by the user is broken into five discrete regions of worst value to best value. The middle color is yellow and the color immediately above it is chartreuse, while the color immediately below yellow is orange. Note that there should be consistency between the scales for goals and colors. Figure A.18 shows the part of the scorecard worksheet that is used to set the color values.

Scorecard Body—The Utility Values. The body of the scorecard is the intersection of the option rows and the measures. It must be filled in by judgment, values from other scorecards, models, or functions attached to worksheet cells. Models and scorecards might be directly linked to the cells in the body of the scorecard or the values may be compiled from decoupled analysis. Filling this information in is often a large analytic task, and DynaRank cannot do this part automatically. We have found it frequently useful to link a model directly to the scorecard to reflect immediately the effects of changes in model parameters, assumptions, etc. We will illustrate this later in the appendix.

The Aggregate Utility Column. Figure A.19 shows this column. The column contains the aggregation of an option's value across all criteria, appropriately weighted. This is the value that is used in cost-effectiveness calculations and cumulative effectiveness estimation. It is where the rollup of how the option scores on the weighted measures is shown.

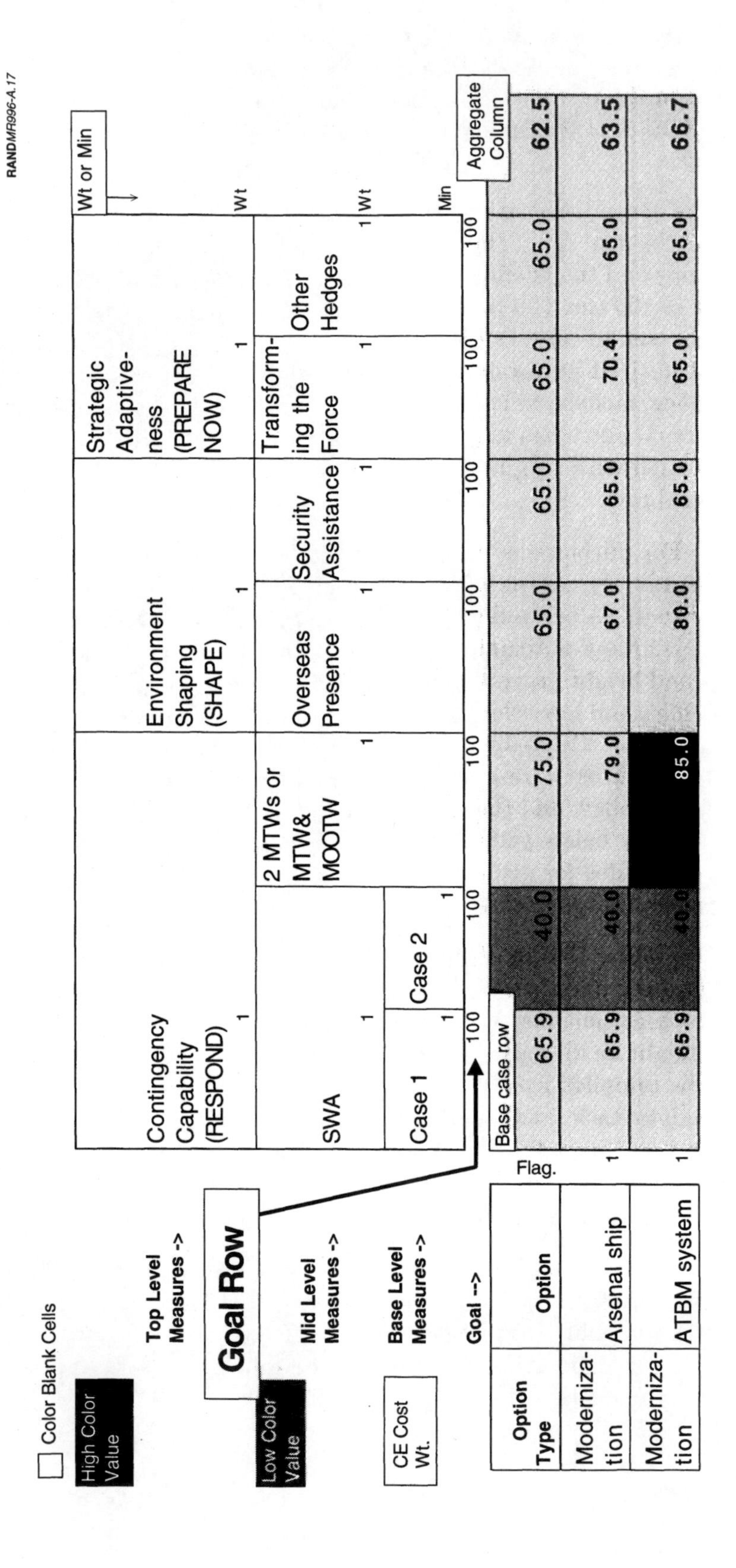

Figure A.17—The Goal Row of a Scorecard

RANDMR996-A.18

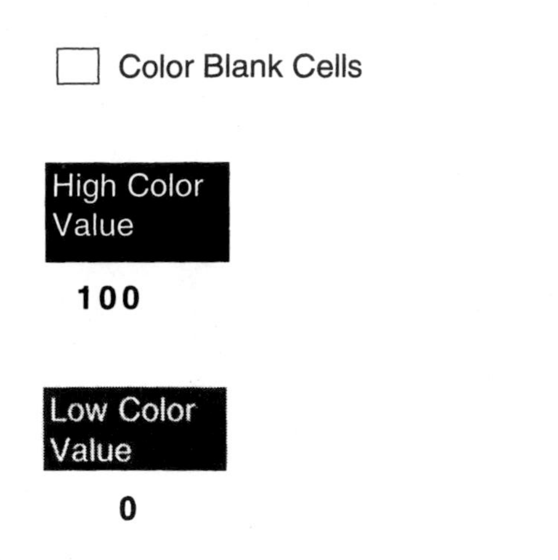

Figure A.18—Color Values
(See Plate VII for color illustration)

The Rank Columns. These are not yet implemented in the production version of DynaRank, but they will show the numerical ranking of the options on cost-effectiveness, effectiveness, and cost. Currently the options are sorted to show these rankings when the Rank function is selected. The Results worksheet, described later, contains these numerical rankings as well.

The Cost and Aggregate Cost Columns. The cost columns appear to the right of the scorecard with one cost measure per column and the cost rows aligned with the options. Figure A.20 shows the cost columns of a DynaRank scorecard. The cost multipliers or weights of the cost components appear immediately below the names of the cost components. These multiply the data in the columns immediately below, and the sum of the weighted columns gives the aggregate or Total Cost column. In contrast to the use of weights in the measures portion of the scorecard, the cost weights are not normalized. So, rather than the columns below being multiplied by 1/3, 1/3, 1/3 as would happen to measures columns, the multipliers are actually 1, 1, 1 as they appear in the Cost Multiplier cells. This was done to allow costs to add up when they represent independent components making up the total cost. Also, it is possible to use discounting by time period if the columns represent costs in different time periods, etc. The Total Cost column is subsequently used to determine cumulative costs and cost-effectiveness. As with the effectiveness, it is assumed that the base case represents the baseline cost. However, the costs are assumed to be the marginal costs of the option, rather than the baseline cost plus the marginal cost of the option. Thus, if option 1 and 2 below were independent, the cost of adding both to the baseline would be 1.4 + 1.5 = 2.9. The units and scale for costs can be whatever the user chooses.

Cost-Effectiveness Columns. These columns carry cost-effectiveness information derived from the aggregate utility of an option and the aggregate cost of the

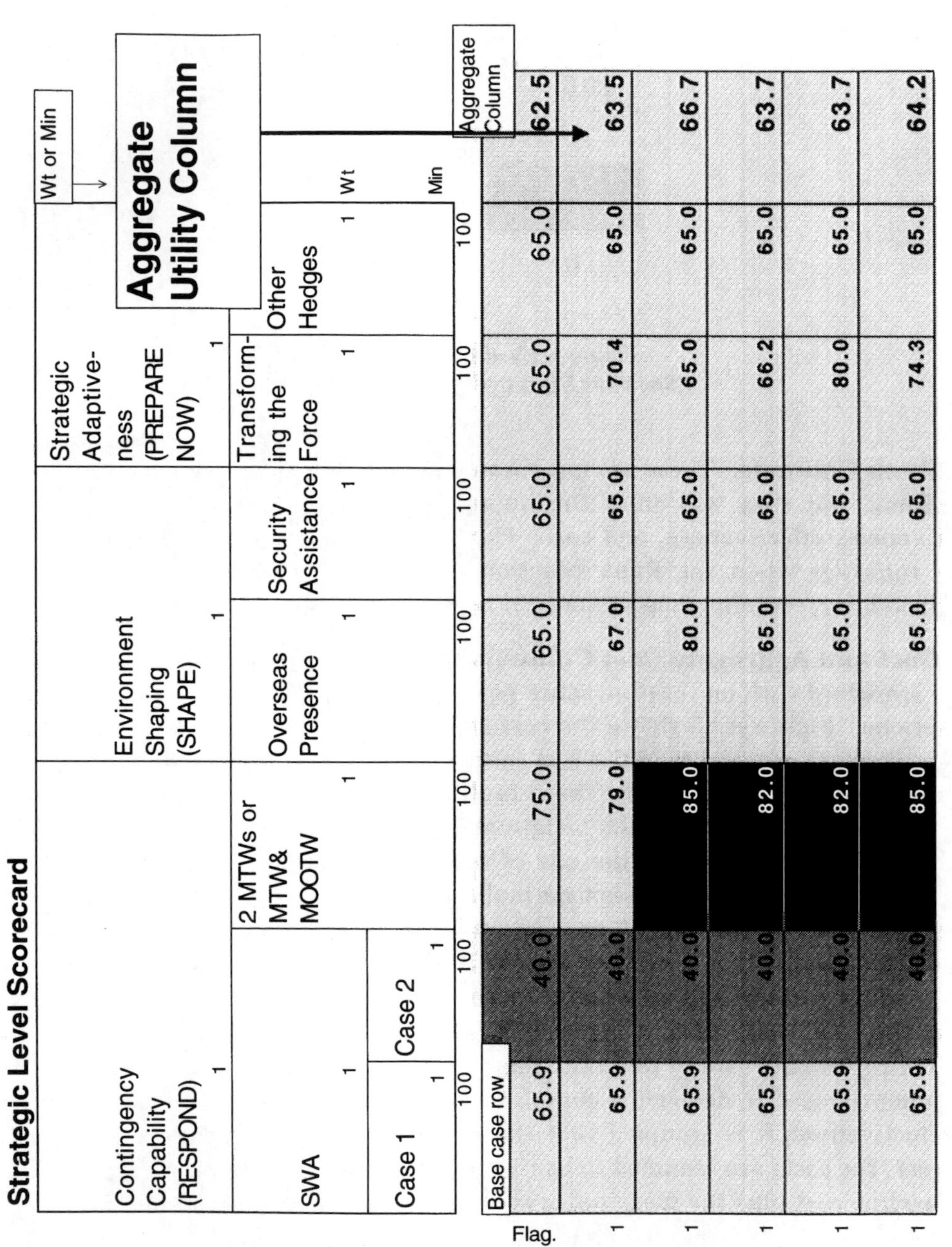

Flag.	Case 1	Case 2	2 MTWs or MTW& MOOTW	Overseas Presence	Security Assistance	Transform-ing the Force	Other Hedges	Aggregate Column
	1	1	1	1	1	1	1	Wt
	100	100	100	100	100	100	100	Min
1	65.9	40.0	75.0	65.0	65.0	65.0	65.0	62.5
1	65.9	40.0	79.0	67.0	65.0	70.4	65.0	63.5
1	65.9	40.0	85.0	80.0	65.0	65.0	65.0	66.7
1	65.9	40.0	82.0	65.0	65.0	66.2	65.0	63.7
1	65.9	40.0	82.0	65.0	65.0	80.0	65.0	63.7
1	65.9	40.0	85.0	65.0	65.0	74.3	65.0	64.2

Figure A.19—The Aggregate Utility Column

RANDMR996-A.20

Costs

Wt	Research and Develop-ment	Acquisition	Operations and Support	Total Cost	Option
	1	1	1		
Base case row	**0.5**	**0.6**	0.3	1.4	
	0.2	0.8	0.5	1.5	Arsenal ship
	0.3	0.9	0.5	1.7	ATBM system
	0.4	0.9	0.5	1.8	Equip F-22s for anti-armor missions
	0.6	1.0	0.4	2.0	C4ISR package

Figure A.20—The Cost Columns of the DynaRank Worksheet

same option and also show the cumulative values of cost and effectiveness. Figure A.21 shows the cost-effectiveness portion of the DynaRank scorecard. The cost-effectiveness type, C-E Type, depends on the sign of the marginal change in cost and effectiveness of an option. Type 1 includes options that show a marginal improvement in aggregate utility and a marginal savings in costs—a win-win option. The cost-effectiveness value for this type of option is the product of the negative of cost (that is, savings) and the marginal improvement in effectiveness. When options are ranked by cost-effectiveness, this type is ranked first. Type 2 cost-effectiveness corresponds to those options that show savings and no effectiveness improvement. The cost-effectiveness value for this type is just the negative of the cost value. Type 3 cost-effectiveness includes those options that have effectiveness improvement and no cost. The cost-effectiveness value in this case is the marginal improvement in effectiveness. Type 4 options are those with improvement in effectiveness and a marginal positive cost. The cost-effectiveness value for these is the ratio of effectiveness to cost. Type 5 options are those for

RANDMR996-A.21

Cost Effectiveness

Option	C-E Value	C-E Type	Cum Eff	Cum Cost
Arsenal ship	7.57	4.00	66.56	0.25
ATBM system	4.17	4.00	70.72	1.25
Equip F-22s for anti-armor missions	27.32	4.00	72.09	1.30
C4ISR package	9.17	4.00	75.76	1.70
Rapid SEAD (HPM)	29.27	4.00	78.98	1.81
Standoff munitions	13.39	4.00	81.66	2.01
Allied defense package (helos/ ATACMS/ advisors)	30.89	4.00	90.92	2.31

Figure A.21—The Cost-Effectiveness Columns

which effectiveness decreases but there is a cost savings. The cost-effectiveness value for this type is the ratio of the negative of the cost to the negative of the effectiveness decrease. The last category of options contains those for which effectiveness decreases and costs increase. This not-very-promising category has the cost-effectiveness computed using 1.0 divided by the product of the cost increase and the negative of the effectiveness decrease. This category is always ranked last except for options excluded by setting the Inclusion Flag to 0.

The cumulative cost and cumulative effectiveness columns are derived from the cost and effectiveness parts of the scorecard. As described earlier, the cumulative effectiveness is computed assuming that an option's marginal aggregate effectiveness can be computed by subtracting the base case from the aggregate effectiveness associated with the option. The marginal cost is transferred directly from the total cost associated with the option. There are applications of the scorecard that do not utilize cost or cost-effectiveness information and these columns should just be ignored in such cases. It is also generally true that costs and effectiveness are not additive, as is implied by the cumulative columns. In cases where the effectiveness and costs are truly marginal values, these cumulative values may be good approximations. In these instances, the user can at least get an approximate, first-order idea of the marginal contribution of a number of options taken together. When options are sorted by cost-effectiveness, they are first sorted by cost-effectiveness type in increasing order, so that all type 1 options appear first, type 2 options appear second, etc. Within a category they are sorted by decreasing cost-effectiveness value.

Derived Scorecards for Top-Level and Level 2 Measures. A DynaRank scorecard worksheet contains two additional scorecards derived from the main scorecard with its hierarchical measures. These secondary scorecards show a partial rollup or aggregation of the measures. The first derived scorecard, appearing immediately to the right of the cost-effectiveness columns, shows the aggregation of performance of the options to each of the top-level measures. Thus, in the example scorecard used here, the performance of the options is shown for Contingency Capability, Environment Shaping, and Strategic Adaptiveness. This aggregation is, of course, dependent on the weights given at the lower levels in the primary scorecard. This rollup is illustrated in Figure A.22.

The second derived scorecard rolls the measures up to the level 2 measures based on the weights on the level 3 measures in the primary scorecard. This is illustrated in Figure A.23. The value in these derived scorecards changes whenever a user changes the weights or measures in the primary scorecard.

Important Assumptions

It is important to know that several assumptions about the meaning of scorecard values and operations have been made:

Normalized Utility Values. The values in the scorecard associated with the various measures for evaluating an option are assumed to be of the same scale and imply increasing utility as the value increases. This permits additivity and the normalization of weights on a scale from 0–1 without any further transformation.

Defaults for Goals. Goals are assumed to be represented in the goal row, and the values range from 0 to 100 by default. If a different scale is desired, this scale is entered in the color values section of the scorecard, then all values in the scorecard should be entered to be consistent with this scale.

Use and Nonuse of Base-Case Values. The base-case values are used in the determination of marginal improvements due to an option. If the marginal improvements are entered in the scorecard directly, then by setting the base-case values to zero, the cumulative effectiveness can still be applied. However, it will not be completed correctly if there is a base-case value and only the marginal improvements in effectiveness are entered in the scorecard under the criteria.

Assumptions About Costs and Marginal Costs. Costs *are* assumed to be the marginal costs so that there is no subtraction of a base-case cost from the costs of

RANDMR996-A.22

Scorecard-Aggregation to Top Level Measures

Lvl1 Wts.	1	1	1
Option	Contingency Capability (RESPOND)	Environment Shaping (SHAPE)	Strategic Adaptiveness (PREPARE NOW)
BaseCase	64.0	65.0	65.0
Arsenal ship	66.0	66.0	67.7
ATBM system	69.0	72.5	65.0
Equip F-22s for anti-armor missions	67.5	65.0	65.6
C4ISR package	67.5	65.0	72.5
Rapid SEAD (HPM)	69.0	65.0	69.7
Standoff munitions	68.1	65.0	68.9
Allied defense package (helos/ ATACMS/ advisors)	74.3	75.0	72.5

Figure A.22—Aggregate Scorecard for Top-Level Measures (See Plate VIII for color illustration)

options. The cost weights are not normalized so that the aggregate costs are the sum of the shown cost multipliers times the cost components.

Use of Equations in the Scorecard. The computed values in the scorecard are obtained by placing equations in those cells. The user should not attempt to directly change the values in the aggregate utility column, aggregate cost column, all columns of the cost-effectiveness columns, and all columns of the aggregate scorecards. Doing so will erase the aggregation equations so that no further computations will be made automatically in those cells.

RANDMR996-A.23

Scorecard-Aggregation to Mid Level Measures

Lvl2 Wts.	1	1	1	1	1	1
	Contingency Capability (RESPOND)		Environ-ment Shaping (SHAPE)		Strategic Adaptive-ness (PREPARE NOW)	
Option	SWA	2 MTWs or MTW & MOOTW	Overseas Presence	Security Assistance	Transform-ing the Force	Other Hedges
BaseCase	52.9	75.0	65.0	65.0	65.0	65.0
Arsenal ship	52.9	79.0	67.0	65.0	70.4	65.0
ATBM system	52.9	85.0	80.0	65.0	65.0	65.0
Equip F-22s for anti-armor missions	52.9	82.0	65.0	65.0	66.2	65.0
C4ISR package	52.9	82.0	65.0	65.0	80.0	65.0
Rapid SEAD (HPM)	52.9	85.0	65.0	65.0	74.3	65.0
Standoff munitions	56.2	80.0	65.0	65.0	72.8	65.0
Allied defense package (helos/ ATACMS/ advisors)	53.5	95.0	75.0	75.0	80.0	65.0

Figure A.23—Derived Scorecard for Rollup to Level 2 Measures (See Plate IX for color illustration)

BASIC USE OF A SCORECARD—COLORING, WEIGHTING, RANKING

Coloring and Uncoloring the Scorecards

Perhaps the simplest operation to perform on the scorecard is to color the cells associated with the measures. When the user selects the **ColorScorecard** menu item from the SC Tools menu, the scorecard will be colored with five colors ranging from bright green for very good (highest value) to bright red for very bad (lowest value) with yellow as the middle value and orange and chartreuse for the other values. The highest value and lowest value for this color coding is designated by the user in the color values boxes described above. When the selection to color the scorecard is chosen, the scorecard will temporarily be moved off the screen as the colors are computed and then switched back on with the coloring done. This saves time because otherwise the software interacts with the scorecard one cell at a time.

The **UncolorScorecard** menu item removes the colors in the scorecard. It is not necessary to uncolor before recoloring a scorecard to reflect new computed or input values. The secondary scorecards and aggregate value columns are all colored at the same time as the main scorecard. *The user should beware that coloring is one function that is not done automatically on a scorecard because of software limitations.*

Weighting Measures and Determining Aggregate Effectiveness

We have already discussed the use of weights and the computation of aggregate effectiveness. Any time a weight is changed, the aggregate effectiveness is recomputed automatically by the equations in the aggregate utility columns. The cost-effectiveness columns and the derived scorecards are also automatically computed. The scorecard is not automatically colored and options are not automatically ranked when these changes occur. The discussion above describes how the weights are normalized at the various measures levels. Changing **Wt** in the aggregate utility column to **Min** at any level causes the worst case for that level to be rolled into the next higher level rather than the weighted sum.

Ranking Options by Effectiveness, Cost, and Cost-Effectiveness, and Unranking

The options can be reordered in rank order of decreasing effectiveness, increasing cost, or decreasing cost-effectiveness by the selection of the appropriate command in the SC Tools menu, as shown in Figure A.24. The ranking of options causes the scorecard rows to be ordered in rank order with the highest ranked at the top. This in turn causes the cumulative effectiveness and costs to be recomputed to reflect the accumulation according to the new order. If the scorecard was colored with the current weights and values, then the colors, which move with the rows, remain valid. The cumulative charts, to be discussed shortly, are based on the new ordering of options and resulting accumulations. The **UnRank** command causes the scorecard options to be placed in the original order of the numbering of the options (the order that options were entered in the scorecard template).

Plotting Cumulative Costs and Effectiveness

It is sometimes useful to see the cumulative buildup of cost and effectiveness as options are added to the base case in rank order. This can be done by choosing the automatically generated graphs from the SC Tools menu. The graph can be created for the aggregate total effectiveness and the total cost, or it can be made with one cumulative effectiveness for each of the top-level measures. Figure A.25 shows the selection of the aggregate graph, and Figure A.26 shows the resulting graph for our example scorecard.

The cumulative chart shows the buildup of cost and effectiveness from the base case as the options are added in current rank order (the order appearing in the scorecard) from left to right. Effectiveness is capped at the aggregate goal effec-

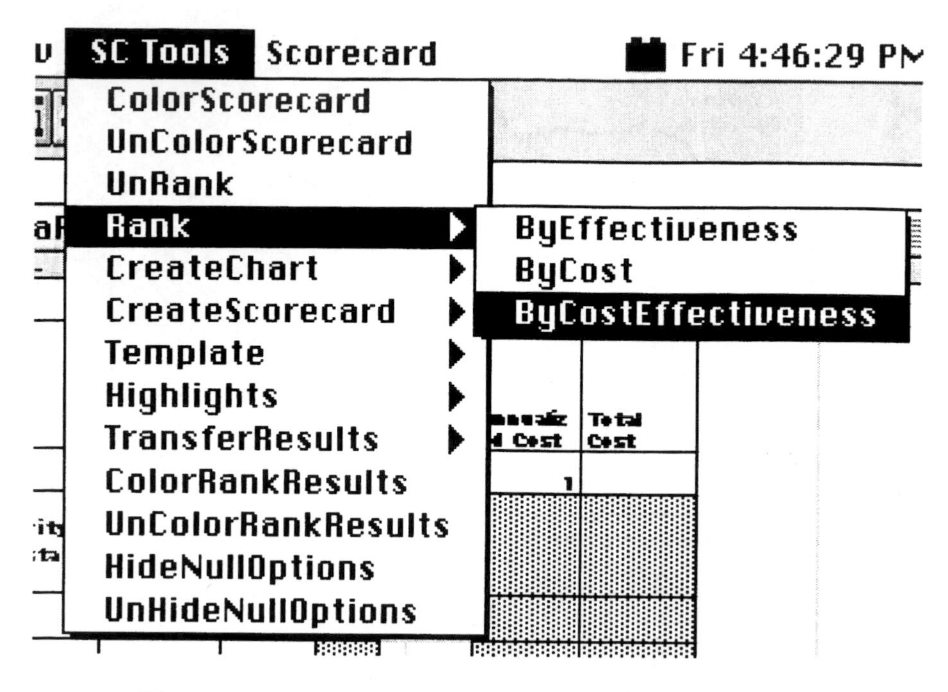

Figure A.24—Choosing to Rank the Scorecard Options by Cost-Effectiveness

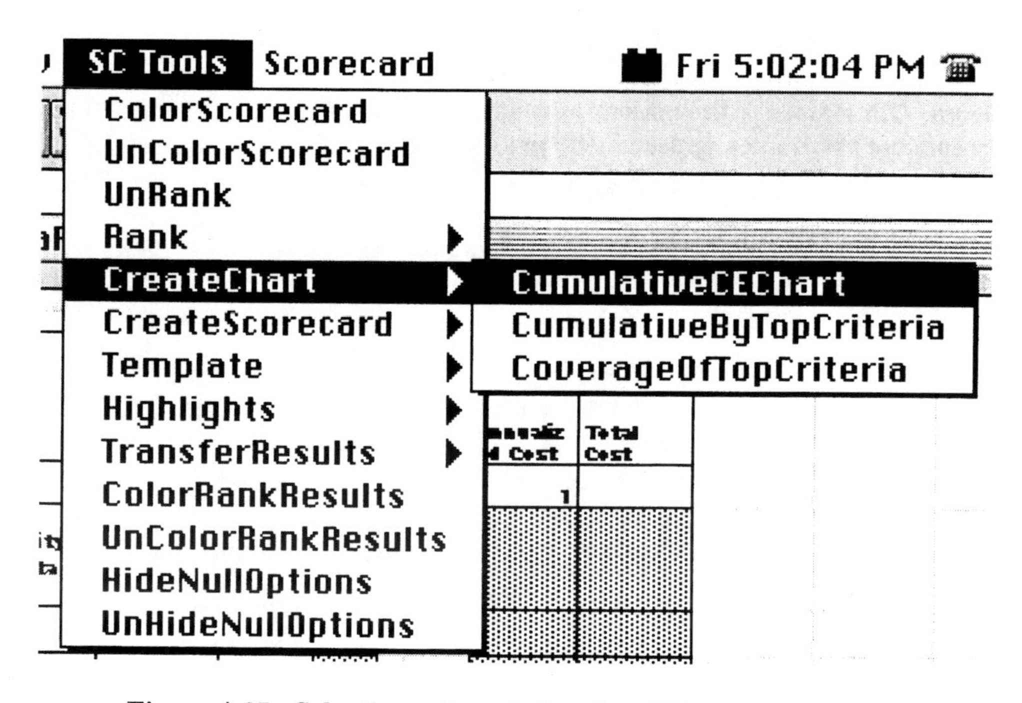

Figure A.25—Selecting a Cumulative Cost-Effectiveness Chart

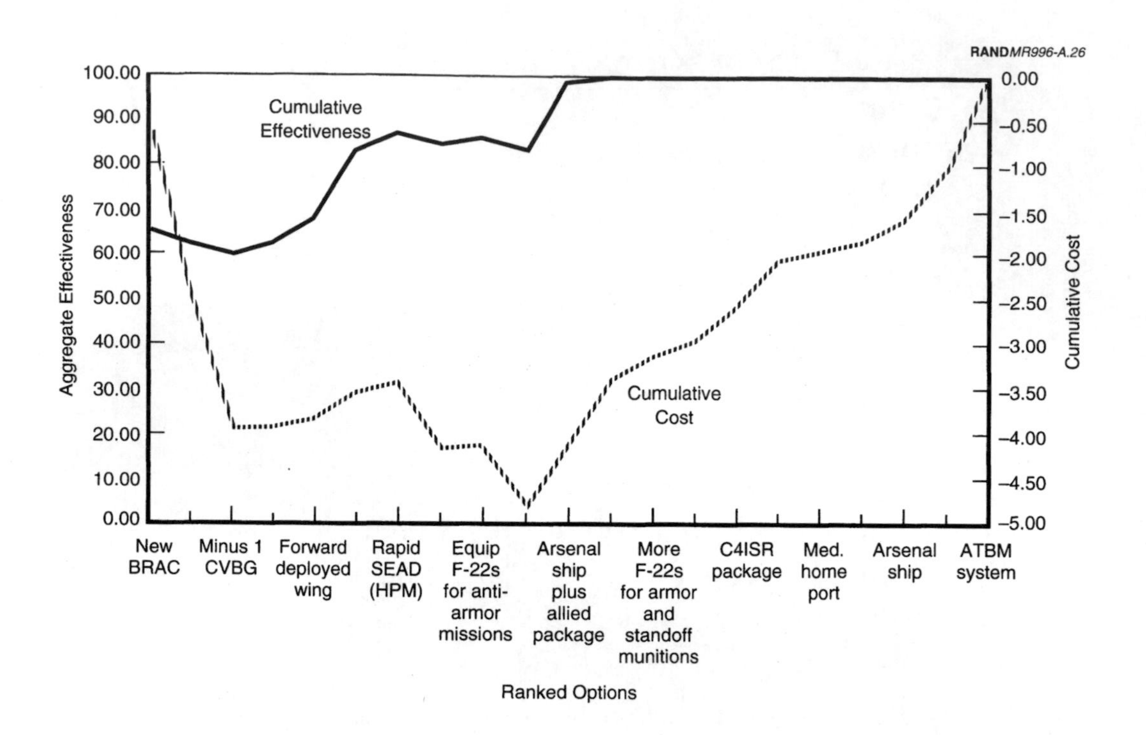

Figure A.26—Cumulative Cost and Effectiveness Chart

tiveness. This chart will be automatically updated whenever the options are reordered. The names of the options appear on the absciss of the graph, but usually only some of the names appear if the list of options is very long or the names are long. The names will eventually show up if the graph is stretched enough by dragging one of its corners. An alternative is to use either the graph inclusion column on the scoreboard to limit certain options from appearing, or to use a number other than the default **All** in the dialogue box that requests the number of options to graph when the chart is created. For example, the same graph, limited to the top 10 options in the ranked list, would look like that shown in Figure A.27.

A similar chart to that in Figure A.26 is obtained when the alternative chart selection, **CumulativebyTopCriteria,** is used except that the cumulative effectiveness for each of the top-level criteria is added to the aggregate cumulative effectiveness and cost. This chart is shown in Figure A.28. This chart is useful in identifying how the performance on each of the measures changes as the options are added in rank order.

There are some aspects of the rank ordering of cumulative effectiveness that should be made clear. The fact that options are first sorted by cost-effectiveness type means that the options cannot move outside of their category order on the graph. Thus, options that save money and improve effectiveness will always appear first. It is possible to change the relative importance of options in category 4 (those for which there is effectiveness improvement and an increase in cost and the cost-effectiveness value is a ratio of the effectiveness divided by the cost) by chang-

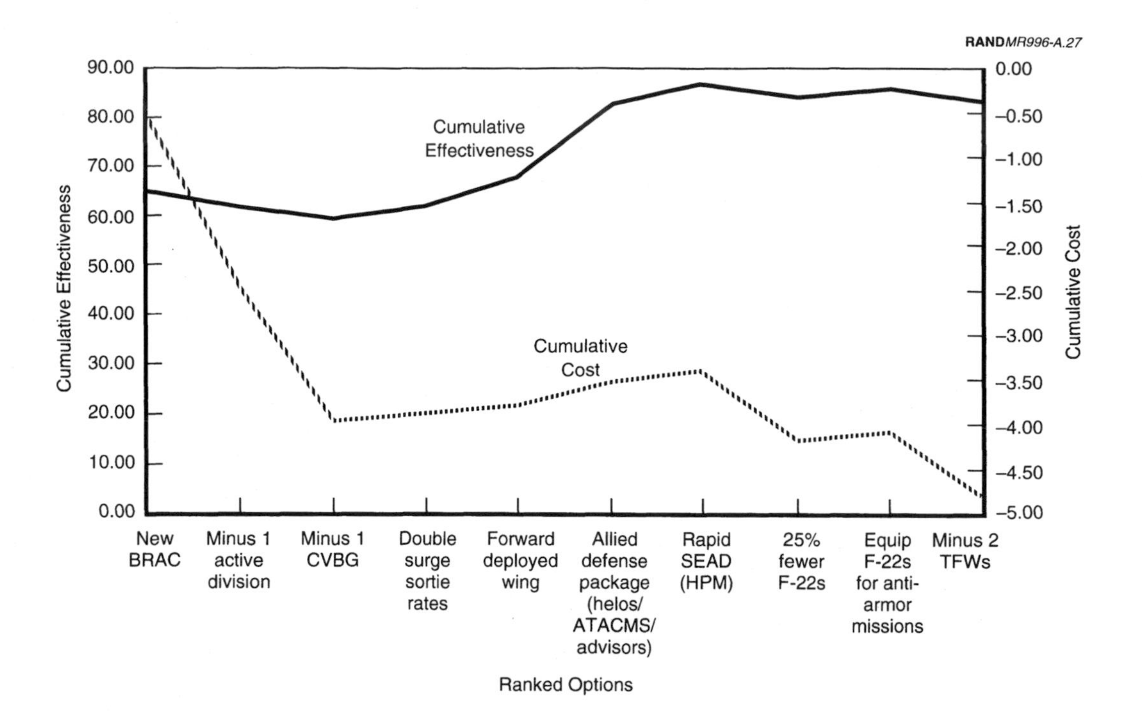

Figure A.27—The Cumulative Cost and Effectiveness Chart Limited to the Top 10 Ranked Options

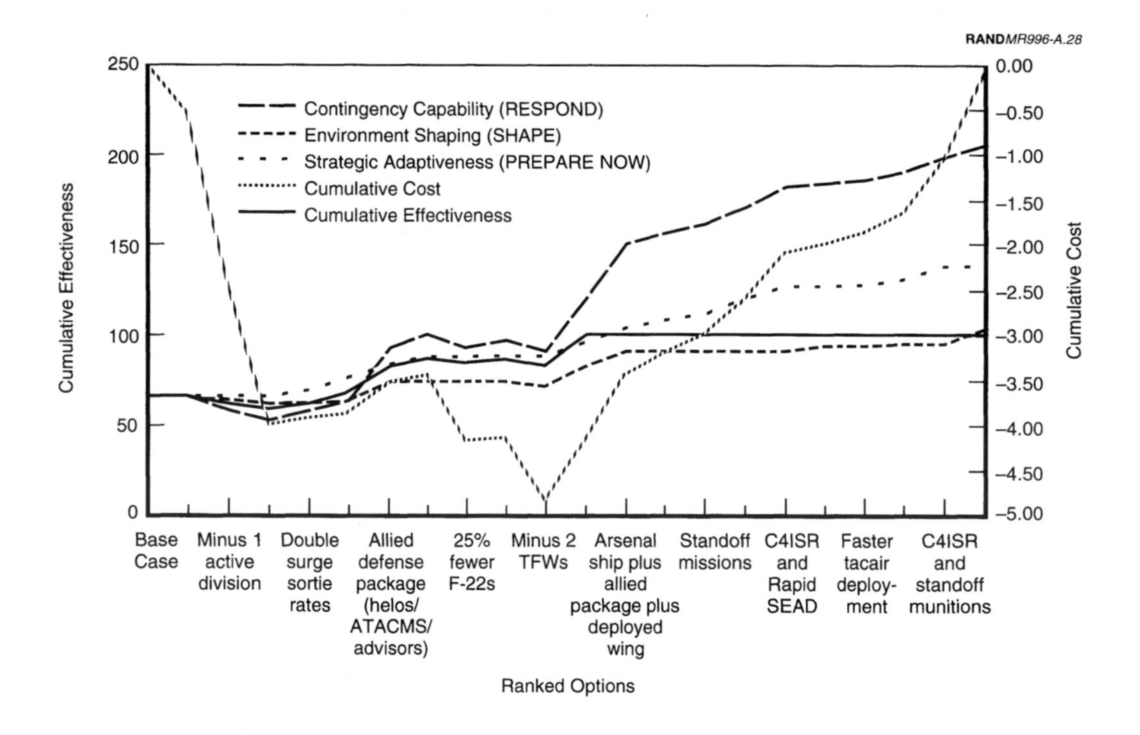

Figure A.28—Cumulative Effectiveness by Top-Level Measures and Cost

ing the weight on cost in the upper left-hand part of the scorecard worksheet. This weight, nominally set to one, causes a change in the relative importance of cost and effectiveness in the ranking of the options. A high weight on cost will cause DynaRank to attempt to choose those effectiveness improvements that do not cost very much and a low weight on cost will cause DynaRank to choose those options with the most improvement in effectiveness. Theoretically, if there was a budget target, this cost weight would be used in an iterative manner to find the most improvement in effectiveness for a given budget. The user can apply it iteratively to examine the effect on the ranking of options.

ACCUMULATING RESULTS

Rank Sheets

So far we have discussed the use of the scorecard tools to perform individual operations, such as ranking, plotting, and coloring. We believe that the weights on criteria and cost represent alternative "views" of strategy or criteria. In the case of the transportation scorecards, the weights might represent different stakeholders, such as those concerned with the environment, those more concerned with the economy, and the focus of various government agencies. In the case of defense strategy, the views might represent those who believe we should focus more on current readiness for contingencies, those who believe we should focus more on transforming the force for the future, and those who believe that the environment shaping aspects of our forces are most important. Rather than attempt to get a "correct" or an "average" view, we assume that all of these views might be represented at one time or another, and that the interesting exercise is to look for options that are important across all of the views or to show the consequences of particular views on the importance of options. For purposes of accumulating alternative views, we have created the Rank sheet and Results sheet to store the views and even perform some operations on those views. The simplest of the accumulations is the Rank sheet.

Creating a Rank Sheet. At any point in the use of the scorecard, the user can choose to store a view by choosing the **Rank** submenu item in the **TransferResults** menu item of the SC Tools menu as shown in Figure A.29. This function can only be performed while the user is working on a scorecard sheet. Once this function is chosen, the user is first asked for the name of the Rank sheet to which the results should be transferred. Either a new sheet can be selected or an existing Rank sheet can be used. After giving the name of the Rank sheet, the user is asked for the name of the rank or view. This becomes a label for the Rank sheet column that contains this particular view and is used to later relay the source of the different views. Figure A.30 shows a Rank sheet with a single view.

Storing Additional Results in a Rank Sheet. Generally an analyst is interested in transferring multiple views to a Rank sheet. This is done simply by choosing the TransferResults and Rank menu items again, giving the name of the previously used Rank sheet and then giving the name for the new results. Figure A.31 shows a Rank sheet with three different views.

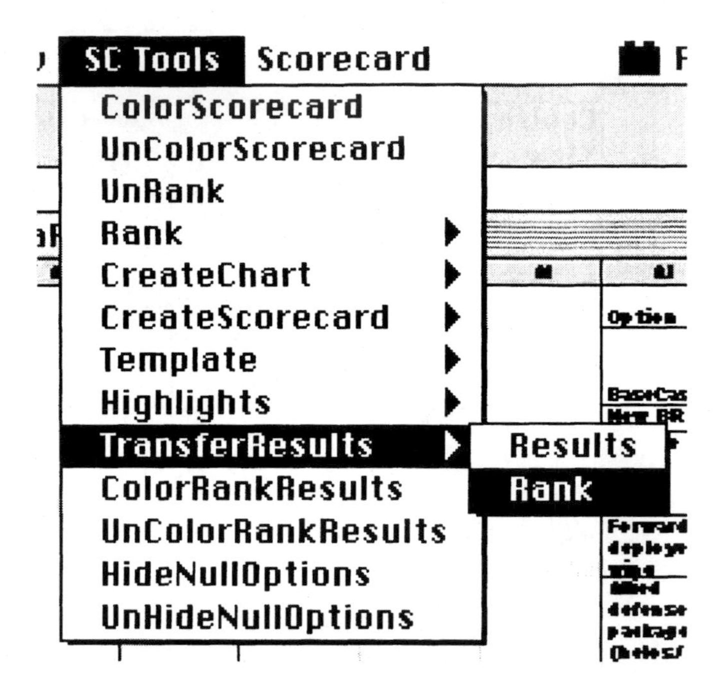

Figure A.29—Choosing to Transfer Results to a Rank Sheet

Count of Ranks	Number for Top x
1	1
	New Rank
1	New BRAC
2	Double surge sortie rates
3	Forward deployed wing
4	Allied defense package (helos/ ATACMS/ advisors)
5	Rapid SEAD (HPM)
6	Equip F-22s for anti-armor missions
7	Arsenal ship plus allied package
	Arsenal

Figure A.30—A Rank Sheet with One Transferred View Called "New Rank"

Count of Ranks	Number for Top x		
3	10		
	Contingency View	Shaping View	Adapativeness View
1	New BRAC	25% fewer F-22s	Minus 1 active division
2	Double surge sortie rates	New BRAC	Minus 1 CVBG
3	Allied defense package (helos/ ATACMS/ advisors)	Allied defense package (helos/ ATACMS/ advisors)	25% fewer F-22s
4	Equip F-22s for anti-armor missions	Forward deployed wing	Minus 2 TFWs
5	Rapid SEAD (HPM)	Med. home port	New BRAC
6	Forward deployed wing	Arsenal ship plus allied package	Forward deployed wing
7	Arsenal ship plus allied package	Arsenal ship plus allied package plus deployed wing	Double surge sortie rates
	Arsenal ship		

Figure A.31—A Rank Sheet with Three "Views" Representing Different Emphases

Coloring and Identifying Top Ranked Options. To help with identifying the common items in these views, the user can elect to color common options in the top x (where x is chosen by the user in the block shown to be 10 in this example) of all the views. This coloring is illustrated in Figure A.32 and is done by selecting the menu item **ColorRankSheet** in the SC Tools menu. The coloring can be removed by choosing the **UnColorRankSheet** menu item.

RANDMR996-A.32

	Count of Ranks	Number for Top x		
	3	10		
24		Contingency View	Shaping View	Adaptiveness View
	1	New BRAC	25% fewer F-22s	Minus 1 active division
	2	Double surge sortie rates	New BRAC	Minus 1 CVBG
	3	Allied defense package (helos/ ATACMS/advisors)	Allied defense package (helos/ ATACMS/advisors)	25% fewer F-22s
	4	Equip F-22s for anti-armor missions	Forward deployed wing	Minus 2 TFWs
	5	Rapid SEAD (HPM)	Med. home port	New BRAC
	6	Forward deployed wing	Arsenal ship plus allied package	Forward deployed wing
	7	Arsenal ship plus allied package	Arsenal ship plus allied package plus deployed wing	Double surge sortie rates
		Arsenal ship plus allied package plus		

Figure A.32—Coloring the Common Items in the Top x Items of the Views (See Plate X for color illustration)

Results Sheets

The Results sheet provides another way to look at the accumulation of views or results of weighting options. In the Results sheet, the aggregate costs and utility of all options corresponding to alternative sets of weights on cost and criteria are stored, rather than just the ranking of options. The user is then given the capability to weight the views in the Results sheet to obtain ranking based on a composite of the weighted aggregate costs and utility. The cost, effectiveness, and cost-effectiveness ranking for the composite view is provided in the Results sheet.

Creating a Results Sheet. The first step is to create a view by choosing weights of cost and criteria in a DynaRank scorecard. Then, as done with storing views for the Rank sheets, the user selects the **Results** submenu option of the **TransferResults** menu item in the SC tools menu. This is shown in Figure A.33.

After this selection, the user will be asked for a name of the sheet to which the results will be transferred and a name for this particular selection of results or view. If the name given for the sheet is new (that is, there is not another worksheet by this name), then a new sheet will be created with only these results appearing on the sheet. If it is the name of an existing Results sheet then the new results or view will be added to other views in that Results sheet. Figure A.34 shows a portion of the Results sheet with only one view transferred. Note that the options are listed in unranked order and there are two columns, one for the aggregate utility and the other for the aggregate cost. These aggregate values were transferred from the scorecard and represent the values associated with the particular weights for the view. Note also that there is a cell in which the user can input a weight on the view. When there are more views, the composite rankings will be dependent on these weights.

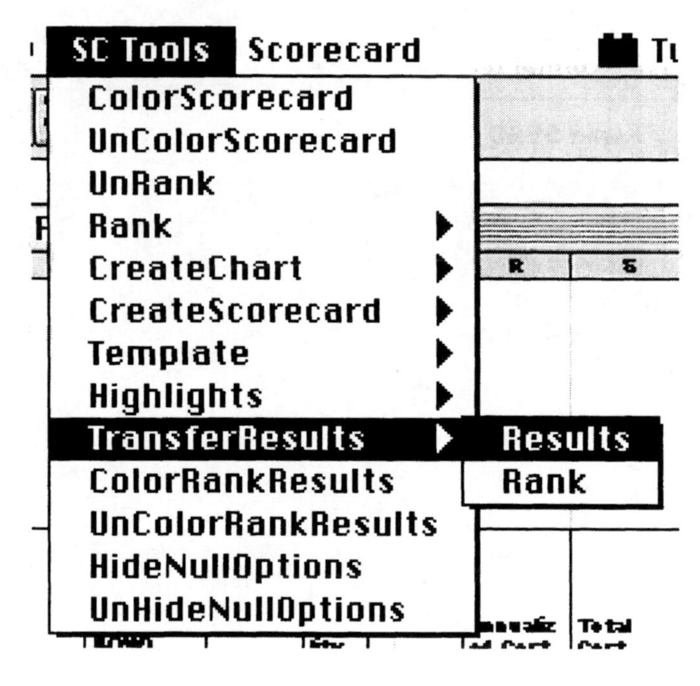

Figure A.33—Transferring Results to a Results Sheet

			Modernization	
			Utility	Cost
Option No.	Option Type	Option Name\Wt	1	
1	Modernization	Arsenal ship	63	0.251
2	Modernization	ATBM system	62	1
3	Modernization	Equip F-22s for anti-armor missions	61	0.05
4	Modernization	C4ISR package	67	0.4
5	Modernization	Rapid SEAD (HPM)	64	0.11
6	Modernization	Standoff munitions	63	0.2
7	Efficiency	Allied defense package (helos/ ATACMS/ advisors)	80	0.3
		Faster		

Figure A.34—A Single View Transferred to a Results Sheet

The rankings are shown in another part of the Results sheet. Figure A.35 shows the composite ranking for this Results sheet in cost, effectiveness, and cost-effectiveness.

Adding Additional Results to the Results Sheet. As described above, additional views can be added to the same results sheet and weighted to create a composite ranking of options. This is done by giving the same Results sheet name and a new name for each view. Three views added to a Results sheet are illustrated in Figure A.36.

Composite Ranking of Options		
Eff. Rank	Cost Rank	C-E Rank
Arsenal ship plus	Minus 1 active	New BRAC
Arsenal ship plus	Minus 1 CVBG	Double surge
Allied defense package (helos/	25% fewer F-22s	Allied defense package (helos/
C4ISR and Rapid SEAD	Minus 2 TFWs	Equip F-22s for
C4ISR and standoff	New BRAC	Rapid SEAD (HPM)
C4ISR package	Equip F-22s for	Forward deployed
Rapid SEAD (HPM)	#N/A	Arsenal ship plus allied package
More F-22s for armor, and standoff	Forward deployed wing	Arsenal ship plus allied
Forward deployed wing	Med. home port	More F-22s for armor, and standoff munitions

Figure A.35—The Composite Effectiveness, Cost, and Cost-Effectiveness Ranking of Options in a Results Sheet

Note that there are now three user weights to be input. As with the scorecard measures these weights are normalized so that they sum to unity. As shown, the normalized weighting of these columns would be 1/3, 1/3, 1/3. These weights then multiply the cost and utility columns and are summed to obtain an aggregate utility and cost of each option across these views. This aggregate is then used to rank options across the views. This ranking is shown in Figure A.37

Using a Results Sheet. The difference between the Results and the Rank sheets is that in the Rank sheet the user can scan a number of rankings to find commonalty among the higher-ranked options. In the Results sheet, there is only one composite ranking, but the user can emphasize one view more than another in obtaining this composite ranking. It deals with the question: Which options are bet-

		Modernization		Shaping		Respond	
		Utility	Cost	Utility	Cost	Utility	Cost
Option Type	Option Name\Wt	1		1		1	
Modernization	Arsenal ship	63	0.251	64	0.251	67	0.251
Modernization	ATBM system	62	1	66	1	70	1
Modernization	Equip F-22s for anti-armor missions	61	0.05	62	0.05	67	0.05
Modernization	C4ISR package	67	0.4	67	0.4	70	0.4
Modernization	Rapid SEAD (HPM)	64	0.11	64	0.11	69	0.11
Modernization	Standoff munitions	63	0.2	64	0.2	67	0.2
Efficiency	Allied defense package (helos/ ATACMS/ advisors)	80	0.3	79	0.3	83	0.3
Efficiency	Faster tacair deployment	61	0.105	63	0.105	66	0.105

Figure A.36—Several Views Transferred to the Same Results Sheet

Composite Ranking of Options		
Eff. Rank	Cost Rank	C-E Rank
Arsenal ship plus	Minus 1 active	New BRAC
Arsenal ship plus	Minus 1 CVBG	Double surge
Allied defense package (helos/	25% fewer F-22s	Forward deployed wing
C4ISR and Rapid SEAD	Minus 2 TFWs	Allied defense
C4ISR and standoff	New BRAC	Rapid SEAD (HPM)
C4ISR package	Equip F-22s for	Equip F-22s for
		Arsenal

Figure A.37—Options Ranked Based on a Composite Weighting of the Views in the Results Sheet

ter on average, rather than Which options appear most often with high rankings? There should be correlation between such options, but not always. The best average option may not rank high on any list.

ADVANCED FEATURES—HIERARCHICAL AND COMBINED SCORECARDS, RISK BARS, ETC.

Building Hierarchical Scorecards

It is quite easy to link DynaRank scorecards into a hierarchical structure. First, the options in each scorecard should be identical, with only the measures changed. To start, the lists should be in the same order. When this is complete, the user then links the scorecards by selecting the aggregate utility column of the more detailed or lower-level scorecard and then pasting the cell references in the appropriate column of the more aggregate or higher-level scorecard. After this is done, the reordering of options in either scorecard will not affect the validity of the linking because the cell references are moved as the scorecards get reordered for ranking, etc. If the lists of options are not the same, but it is desired to link some cells of one scorecard to another for some specific options, then just those cell references would be transferred. Sometimes it is desirable to link results of level 2 or level 3 aggregations of a lower-level scorecard to a higher-level scorecard. This can be done by linking the appropriate columns of the subsidiary scorecards. This was done in the scorecards shown earlier in the appendix. The subsidiary results for case 1 and case 2 of the SWA scenario in the SWA scorecard were linked to the case 1 and case 2 of the Strategic Scorecard. Note that this hierarchical structure could go many scorecards deep by linking additional scorecards in a hierarchical manner. Or multiple detailed scorecards could be linked to the same aggregate scorecard.

Building Combined Scorecards

One tool in DynaRank assists the creation of a composite scorecard from multiple other scorecards. The measures columns of a composite scorecard are obtained from the aggregate utility columns of the supporting scorecards. These can then be weighted to obtain a new aggregate utility column for the composite scorecard. In contrast to the Results sheet, this is a full-fledged scorecard with all the functionality of any other scorecards. Also, when the results from the other scorecards change, the appropriate columns of the composite scorecard change as well. Figure A.38 shows the selection of the **Combined** submenu item in the **Create-Scorecard** menu item in the SC Tools menu.

After this selection, a dialogue box will show the user the current scorecards in the workbook and request that a selection of one or more of these be made by highlighting the ones desired. This is illustrated in Figure A.39. Once the user selects OK in the dialogue box, the combined scorecard will be created with appropriate links to all of the selected scorecards.

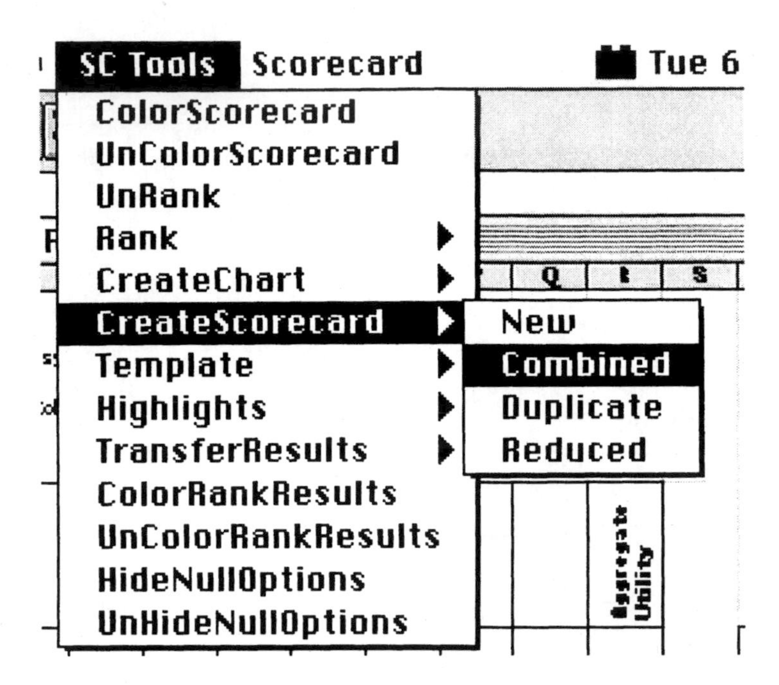

Figure A.38—Creating a Combined Scorecard

The option list for each scorecard must be identical and the options must, at the time the combined scorecard is created, all be in the same order in each scorecard. This is checked by the DynaRank processing, and if not true there will be an error message and the process will abort. Figure A.40 shows the combined scorecard for the three scorecards selected in the dialogue box in Figure A.39.

Note that there is one measure column and one cost column on this scorecard for each of the three scorecards selected in the dialogue box. These correspond to the aggregate utility and aggregate cost columns of those scorecards. Aggregate cost and utility columns of this combined scorecard are based on the weighting of these other columns. Note also that only top-level measures are shown in this scorecard. The values in the measures columns are imported directly from the supporting scorecards and change as those other scorecards change. Reordering the options in any of the scorecards after creation of the combined scorecard is permitted because the cell references get moved automatically by Excel. This scorecard retains all of the functionality of a DynaRank scorecard.

Showing the Colors of One Scorecard in Another

It is possible to show the colors from one scorecard in the cells of another. This is useful, for instance, when one scorecard shows attributes of a measure that are not represented on the other. We have used this to show the risk of achieving the performance measures on one scorecard when the risk is estimated on another. Of course it is always possible to create a scorecard with the risk as a separate col-

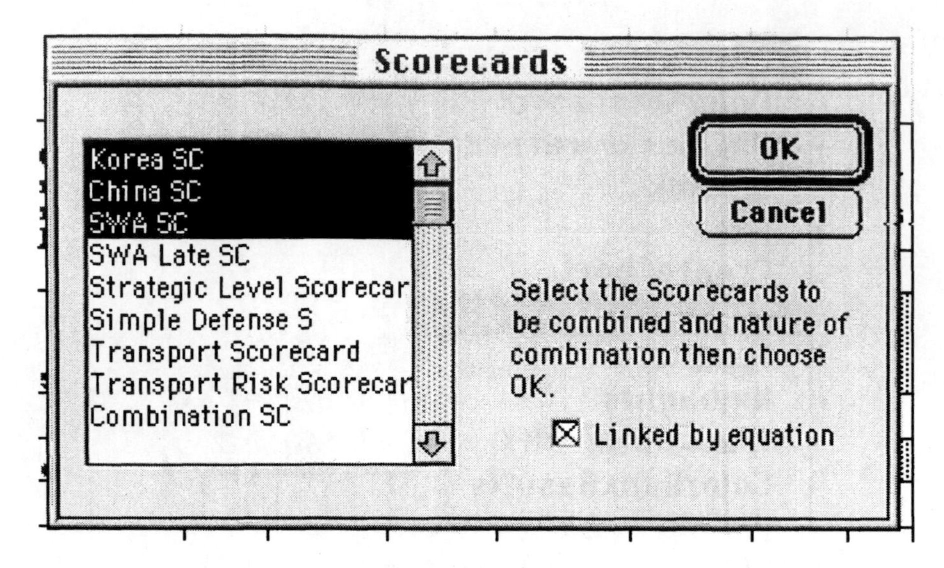

Figure A.39 —Dialogue Box to Create a Combined Scorecard

umn for each measure, but this doubles the size of a scorecard. This feature causes the right and bottom border of a cell to be widened and colored with the transferred colors from another scorecard. The feature is selected by choosing the **Add** submenu item from the **Highlights** menu item of the SC Tools menu, as shown in Figure A.41.

Once this selection is made, the user is provided with a dialogue box that lists scorecards (see Figure A.42). Only one of the scorecards can be selected. DynaRank checks to see if the options are the same and in the same order before applying the highlights. Currently, the size of the body of the scorecard (number of columns and rows) must be the same for this option to work.

The highlights appear as color changes, and a widening of the right and bottom border lines of the corresponding cells occurs. If the colors are the same or there is no color, the borders are not changed. Figure A.43 illustrates the highlighting of cells in this manner. Note that there is text at the top of the scorecard that indicates the source of the highlights. The text and highlights are removed by choosing the **Remove** submenu item of the **Highlights** menu item of the SC Tools menu.

Use of Columns and Weights to Represent Uncertainty

It is possible to use the measures columns and/or cost columns to represent variations in outcomes to which the user can then attach probabilities of those outcomes through the weights. For example, if the cost has significant uncertainty, then

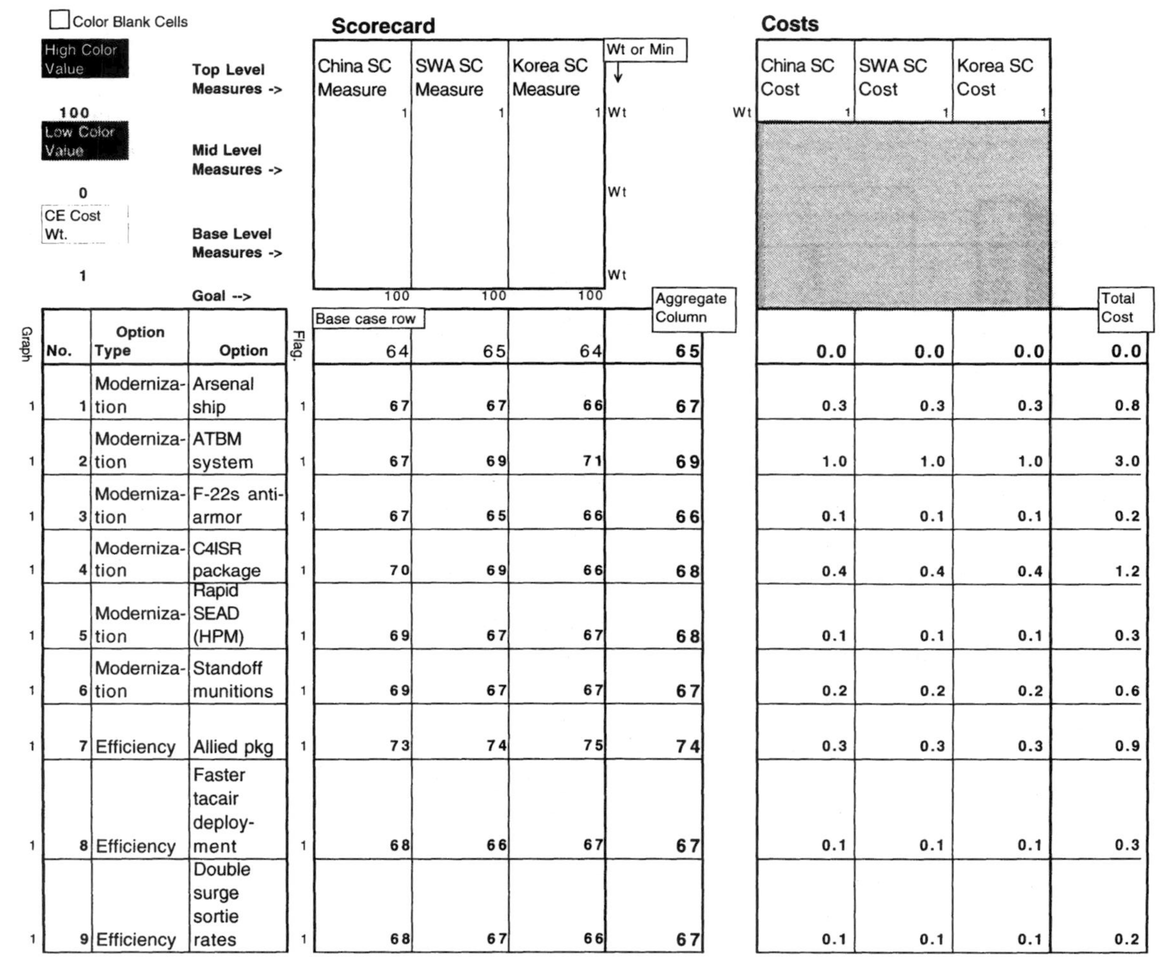

Graph	No.	Option Type	Option	Flag.	China SC Measure	SWA SC Measure	Korea SC Measure	Aggregate Column	China SC Cost	SWA SC Cost	Korea SC Cost	Total Cost
					1	1	1		1	1	1	
					100	100	100					
					64	65	64	65	0.0	0.0	0.0	0.0
1	1	Moderniza-tion	Arsenal ship	1	67	67	66	67	0.3	0.3	0.3	0.8
1	2	Moderniza-tion	ATBM system	1	67	69	71	69	1.0	1.0	1.0	3.0
1	3	Moderniza-tion	F-22s anti-armor	1	67	65	66	66	0.1	0.1	0.1	0.2
1	4	Moderniza-tion	C4ISR package	1	70	69	66	68	0.4	0.4	0.4	1.2
1	5	Moderniza-tion	Rapid SEAD (HPM)	1	69	67	67	68	0.1	0.1	0.1	0.3
1	6	Moderniza-tion	Standoff munitions	1	69	67	67	67	0.2	0.2	0.2	0.6
1	7	Efficiency	Allied pkg	1	73	74	75	74	0.3	0.3	0.3	0.9
1	8	Efficiency	Faster tacair deploy-ment	1	68	66	67	67	0.1	0.1	0.1	0.3
1	9	Efficiency	Double surge sortie rates	1	68	67	66	67	0.1	0.1	0.1	0.2

Figure A.40—A Combined Scorecard

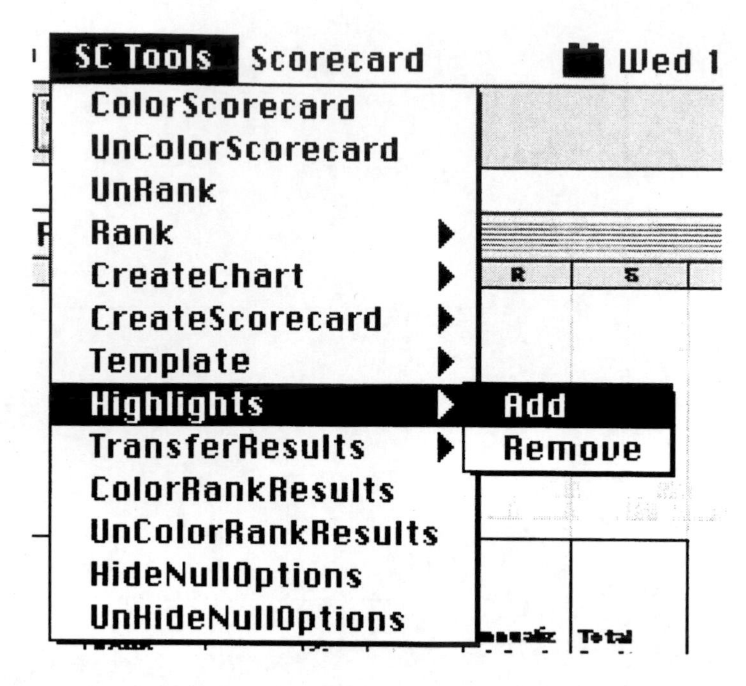

Figure A.41—Choosing to Highlight a Scorecard with Colors from Another Scorecard

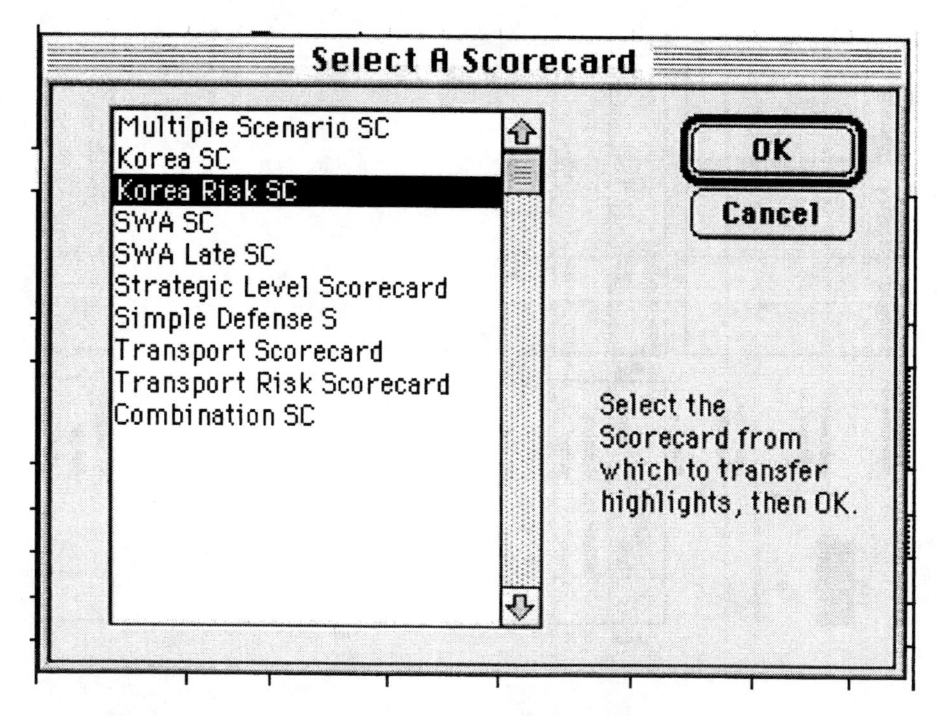

Figure A.42—Choosing the Scorecard from Which to Obtain the Highlight Colors

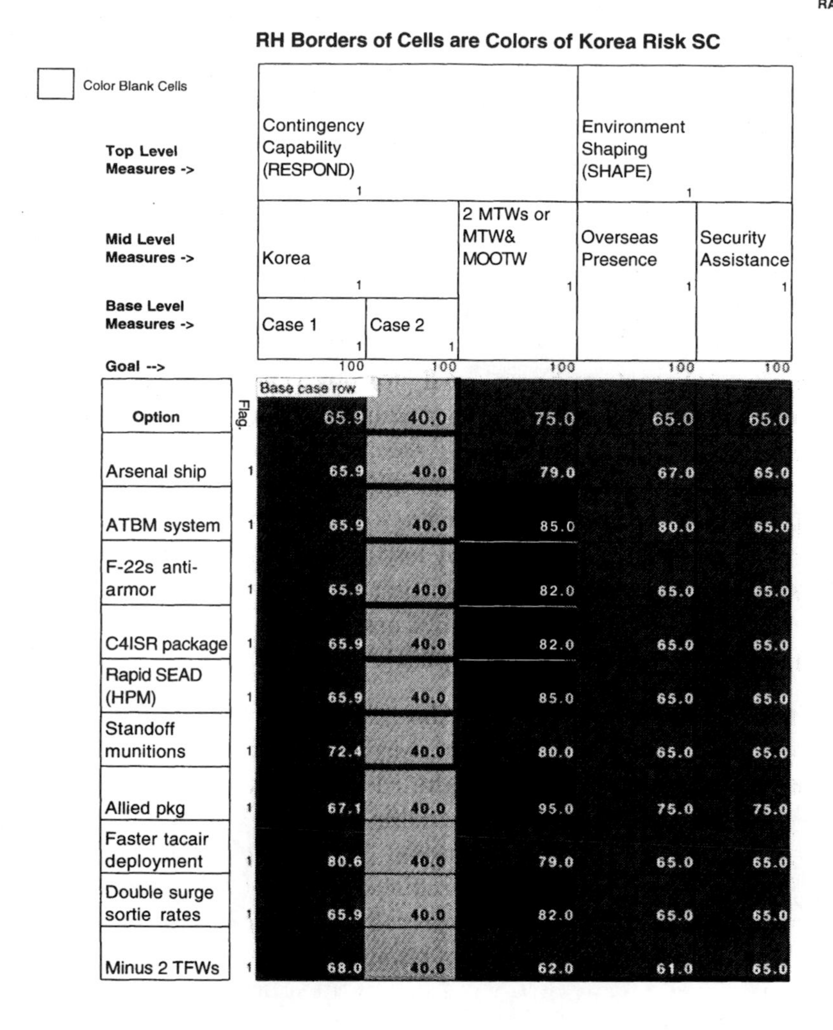

Top Level Measures ->		Contingency Capability (RESPOND) 1			Environment Shaping (SHAPE) 1	
Mid Level Measures ->		Korea 1		2 MTWs or MTW& MOOTW 1	Overseas Presence 1	Security Assistance 1
Base Level Measures ->		Case 1 1	Case 2 1			
Goal -->		100	100	100	100	100
Option	Flag	65.9	40.0	75.0	65.0	65.0
Arsenal ship	1	65.9	40.0	79.0	67.0	65.0
ATBM system	1	65.9	40.0	85.0	80.0	65.0
F-22s anti-armor	1	65.9	40.0	82.0	65.0	65.0
C4ISR package	1	65.9	40.0	82.0	65.0	65.0
Rapid SEAD (HPM)	1	65.9	40.0	85.0	65.0	65.0
Standoff munitions	1	72.4	40.0	80.0	65.0	65.0
Allied pkg	1	67.1	40.0	95.0	75.0	75.0
Faster tacair deployment	1	80.6	40.0	79.0	65.0	65.0
Double surge sortie rates	1	65.9	40.0	82.0	65.0	65.0
Minus 2 TFWs	1	68.0	40.0	62.0	61.0	65.0

Figure A.43—Highlights Shown on a Scorecard (See Plate XI for color illustration)

additional cost columns could be used to represent the various cost possibilities and the weights could be used to represent the probability of that cost outcome. The aggregate cost then represents the expected cost of the option. Similarly, additional columns could be used to represent various scenarios and the weights could be used to represent an estimate of the probability of each scenario. The aggregate utility could then be an estimate of the expected outcome.

Using Linked Models for Utility Values

Just as we illustrated the linking of scorecards in a hierarchy, it is possible to link Excel models directly to outcome measures or costs, so that the individual cells in the scorecard are filled in automatically by models. We have done this, for example, with models of the halt phase of a contingency and models of the presence aspect of the shaping measures. Linking options, models, and scorecards in this way provides a powerful integration methodology for showing the contribution of options in depth as well as at a high level.

A DISCUSSION OF LIMITATIONS

Nonlinear Addition of Effects and Costs of Options

DynaRank provides a calculation of the cumulative effect of options added in order from the top of the list to the bottom. This can be a valid approximation for cases in which the marginal contribution of an option is not large relative to the baseline. However, options do not necessarily (or often) add in the linear manner that such an accumulation assumes. As an example of nonlinear addition, consider options to eliminate highway emissions. One option might make a 50 percent improvement and another, taken by itself, might make a 40 percent improvement. Taken together they are unlikely to make a 90 percent improvement. Often, the combined effect can be approximated by letting the second option affect only the 50 percent remaining, or in other words, letting it have a 40 percent of the 50 percent remaining, or a 20 percent effect. This nonlinear addition can be triggered by checking the box that is labeled **Asymptotic Cum. Eff.** on the scorecard.

In many cases, the cumulative values shown in the scorecard are meaningless and should be ignored. In other cases, they must be assumed to be first-order approximations, and further modeling of combined effects should be performed for those options that appear to be promising combinations. These problems of nonlinear addition also affect the cost accumulations. For example, if the number of a given type of aircraft in a force structure is reduced as a cost-cutting measure and additional capabilities are to be added to aircraft to improve effectiveness, the overall cost of the new capabilities will be less because they are applied to fewer aircraft. At the moment we do not have any options in the scorecard to handle this nonlinear addition of costs.

Nonlinear Weighting of Measures

The use of weights in DynaRank is simple and straightforward, the emphasis being not on getting the weights just right, but on examining the influence of various emphases and views. The user should beware, however, that there is a rich literature dealing with multi-attribute measures and objectives and that this literature discusses approaches to weighting such objectives, obtaining utilities from individuals, and the nonlinear addition of components of objectives. It is also true that the different components of the measures used in DynaRank are not necessarily independent, so that a weight on one might imply a weight on another. We make no claim to mathematical rigor in multi-attribute theory or dealing with dependence among objectives. Rather, this tool is for exploratory and screening analysis of options that ultimately may need to be examined in considerably more depth with respect to outcomes, costs, combinations with other options, etc.

NEXT STEPS—AUTOMATED SEARCHES

Currently the DynaRank system does not automatically search for solutions or automatically perform sensitivity analysis. We expect to add capability, for example, to permit automated variation in weights to determine the range of weights over which an option ranks high. We may also add a capability to find the most effective set of options for which the accumulated costs and savings meet a given budget target.

Appendix B

A LISTING OF DYNARANK FUNCTIONS

There are two DynaRank "pull-down" menus—**SC Tools** and **Scorecard**—the latter lists the current scorecards in the workbook, is recreated each time the workbook is opened, and is updated when a new scorecard is created. It currently does not record scorecard deletions, however, until one of these two actions is performed. When a scorecard name is selected in the pull-down list associated with the **Scorecard** menu item, DynaRank will make that worksheet the active one. The **SC Tools** menu is used to select operations to be performed on a scorecard, template, Results sheet, or Rank sheet. The operations are described below.

SCORECARD FUNCTIONS

Color/Scorecard

This colors the cells of the scorecard including the aggregate utility column and the base-case row using five colors varying from bright red as the lowest value to bright green for the highest value. The middle value is yellow. The colors associated with cell values are determined by the highest color value entry and lowest color value entry shown in cells D3 and E3. Once a scorecard is colored, the colors will follow a cell when it is moved by a ranking or unranking process. Currently colors are not automatically changed as the result of changing cell values or changing weights. You must reselect the **ColorScorecard** command. If **Color Blank Cells** is checked, the color associated with zero will be used. Otherwise blank cells will not be colored. Cells of the secondary scorecards on the same sheet will be colored when the main scorecard is colored.

UnColor/Scorecard

This uncolors all of the colored cells. It is *not* necessary to do this in between coloring and recoloring a scorecard.

UnRank

This places the options and all information in the option rows (including cost-effectiveness tables, secondary scorecard, etc.) in the order that the options were entered in the scorecard template.

Rank

ByEffectiveness. Using the aggregate utility column, this places the option rows in order of highest effectiveness to lowest.

ByCost. Using the total cost column, this places the option rows in order of lowest cost to highest cost.

ByCostEffectiveness. This ranks the options and orders the rows in order of highest cost-effectiveness to lowest within categories of cost-effectiveness type. See the longer discussion of this in Appendix A.

CreateChart

CumulativeCEChart. This creates a graph showing the cumulative cost and cumulative effectiveness as more and more options are added from the highest ranked in the current list to the lowest.

CumulativeByTopCriteria. This creates a graph of options as they contribute to each of the top-level criteria. This chart also shows the contributions to the aggregate effectiveness.

CreateScorecard

New. This creates a new scorecard from information entered in the scorecard template.

Duplicate. This creates an identical copy of the open scorecard worksheet.

Combined. This brings up a list of existing scorecards and, after user selection, causes a new scorecard to be built. Options allow the transfer of the values only or linking to the underlying scorecard using a cell reference. The effectiveness columns of the new scorecard are the aggregate effectiveness columns of the underlying scorecards. The cost columns are the aggregate cost columns of the underlying scorecards. A scorecard template is created and named with the word "template" appended to the new scorecard name. This feature makes it easy to create a hierarchy of scorecards. The options and option order must be the same on each scorecard combined.

Template

Duplicate. This creates a duplicate of the current template (you must be on a template sheet to perform this function).

Clear. This clears the user-entered information from the current template.

New. This creates a new blank template.

Highlights

Add. This requests the name of a second scorecard and then transfers the colors of that scorecard to appear as a thick colored bar at the right side of each cell. The options and order of the options must be the same on each scorecard.

Remove. This removes the transferred colors.

Transfer/Results

Results. This transfers the aggregate utility and the aggregate cost columns to a new worksheet. The name of this worksheet and the name for this set of results is requested. If the name exists, then the columns are added to the Results sheet, otherwise a new Results sheet is created. The Results sheet can be used to accumulate and weight results. Ranking by cost, cost-effectiveness, and effectiveness across weighted results is shown.

Rank. This transfers the current scorecard's list of options to a Rank sheet. The name of this worksheet and the name for this ordering of options is requested. If the name exists, then the columns are added to the existing Rank sheet, otherwise a new Rank sheet is created. The Rank sheet can be used to show which options are robust, that is, which items are common at the top of all lists.

RANK AND RESULTS SHEET FUNCTIONS

ColorRankResults

This is used with a Rank sheet. Options that are common in the top x of all rank lists in the Rank sheet are each the same color. The parameter x is a user input to the Rank sheet.

UnColorRankResults

This uncolors the Rank sheet.

BIBLIOGRAPHY

Air Force Scientific Advisory Board, *New World Vistas: Air and Space Power for the 21st Century*, Department of the Air Force, 1996.

Army TRADOC, *Task Force Griffin: Final Briefing Report,* TRADOC Analysis Center, Fort Leavenworth, Kan., September 1996. Included in Vol. 2 of Defense Science Board, 1996.

Barnett, Jeffery R., *Future War: An Assessment of Aerospace Campaigns in 2010*, Maxwell Air Force Base: Air University Press, 1996.

Bennett, Bruce W., *Two Alternative Views of War in Korea: The North and South Korean Revolutions in Military Affairs,* Santa Monica, Calif.: RAND, MR-613-NA, 1995 (limited to U.S. government agencies only).

Bennett, Bruce W., and Daryl G. Press, *Potential Operational and Strategic Implications of Chemical and Biological Weapons*, Santa Monica, Calif.: RAND, DB-187, forthcoming.

Bennett, Bruce W., Arthur M. Bullock, Daniel B. Fox, Carl Jones, John Schrader, Robert Weissler, and Barry Wilson, *JICM 1.0 Summary*, Santa Monica, Calif.: RAND, MR-383-NA, 1994.

Bennett, Bruce W., Samuel Gardiner, and Daniel B. Fox, "Not Merely Fighting the Last War," in Paul K. Davis, ed., *New Challenges for Defense Planning: Rethinking How Much Is Enough*, Santa Monica, Calif.: RAND, MR-400-RC, 1994.

Betts, Richard, *Surprise Attack: Lessons for Defense Planning*, Washington, D.C.: Brookings Institution, 1982.

Bowie, Christopher J., Frederick L. Frostic, Kevin N. Lewis, John Lund, David A. Ochmanek, and Phil Propper, *The New Calculus: Analyzing Airpower's Changing Role in Joint Theater Campaigns*, Santa Monica, Calif.: RAND, MR-149-AF, 1993.

Carrillo, Manuel J., and Richard Hillestad, *PACE-FORWARD: Policy Analytic and Computational Environment for Dutch Freight Transportation*, Santa Monica, Calif.: RAND, MR-732-EAC/VW, 1996.

Chief of Naval Operations (CNO), *Forward from the Sea*, Department of the Navy, 1997.

Chubin, Shahram, and Charles Tripp, *Iran-Saudi Arabia Relations and Regional Order*, Adelphi Paper 304, London, UK: International Institute of Strategic Studies, 1996.

Cohen, Secretary of Defense William S., *Report of the Quadrennial Defense Review*, Washington, D.C.: Department of Defense, May 1997.

Davis, Paul K., *Toward a Conceptual Framework for Operational Arms Control in Europe's Central Region*, Santa Monica, Calif.: RAND, R-3704-USDP, 1988.

Davis, Paul K., ed., *New Challenges for Defense Planning: Rethinking How Much Is Enough*, Santa Monica, Calif.: RAND, MR-400-RC, 1994.

Davis, Paul K., and Manuel J. Carrillo, *Exploratory Analysis of "the Halt Problem": A Briefing on Methods and Initial Insights*, Santa Monica, Calif.: RAND, DB-232-OSD, 1997.

Davis, Paul K., David C. Gompert, and Richard L. Kugler, *Adaptiveness in National Defense: The Basis of a New Framework*, Santa Monica, Calif.: RAND, IP-155, 1996. Included in Zalmay M. Khalilzad, and David A. Ochmanek (eds.), *Strategic Appraisal 1997: Strategy and Defense Planning for the 21st Century*, Santa Monica, Calif.: RAND, MR-826-AF, 1997.

Davis, Paul K., Richard Hillestad, and Natalie Crawford, "Capabilities for Major Regional Conflicts," in Zalmay M. Khalilzad, and David A. Ochmanek (eds.), *Strategic Appraisal 1997: Strategy and Defense Planning for the 21st Century*, Santa Monica, Calif.: RAND, MR-826-AF, 1997.

Davis, Paul K., and Richard L. Kugler, "New Principles for Force Sizing: Beyond the Two-MRC Criterion," in Zalmay M. Khalilzad, and David A. Ochmanek (eds.), *Strategic Appraisal 1997: Strategy and Defense Planning for the 21st Century*, Santa Monica, Calif.: RAND, MR-826-AF, 1997.

Davis, Paul K., Richard L. Kugler, and Richard Hillestad, *Strategic Issues and Options for the Quadrennial Defense Review (QDR)*, Santa Monica, Calif.: RAND, DB-2010-OSD, 1997. Also see *RAND Research Review*, Vol. XXI, Number 2, Fall 1997.

Defense Science Board, *Investments for 21st Century Military Superiority*, classified report of the 1995 summer study task force, 1995 (limited distribution).

———, *Tactics and Technology for 21st Century Military Superiority*, two volumes, report of the 1996 summer study task force, October 1996.

Department of Defense, *Conduct of the Persian Gulf War*, Final Report to Congress, Washington, D.C., 1992.

Director, Defense Research and Engineering (DDR&E), *Joint Warfare Science and Technology Plan*, Department of Defense, 1996.

Friedman, Norman, *Desert Victory: The War for Kuwait*, Annapolis, Md.: Naval Institute Press, 1991.

Fuller, Graham, and Bruce Pirnie, *Iran, Destabilizing Potential in the Persian Gulf,* Santa Monica, Calif.: RAND, MR-793-OSD, 1996 (limited to U.S. government agencies only).

Goeller, B. F., Allan F. Abrahamse, James H. Bigelow, Joseph G. Bolten, David M. de Ferranti, J. C. DeHaven, T. F. Kirkwood, and Robert L. Petruschell, *Protecting an Estuary from Floods—A Policy Analysis of the Oosterschelde: Vol. I, Summary Report*, Santa Monica, Calif.: RAND, R-2121/1-NETH, 1977.

Goeller, B. F., and the PAWN Team, "Planning the Netherlands' Water Resources," *Interfaces*, Vol. 15, No. 1, 1985, pp. 3–33.

Hillestad, Richard, Warren E. Walker, Manuel J. Carrillo, Joseph G. Bolten, Patricia Twaalfhoven, and Odette Van de Riet, *FORWARD: Freight Options for Road, Water, and Rail for the Dutch: Final Report*, Santa Monica, Calif.: RAND, MR-736-EAC/VW, 1996.

Institute for National Strategic Studies (INSS), *Strategic Assessment, 1995: U.S. Security Challenges in Transition*, Washington, D.C.: National Defense University Press, 1995.

Johnson, Stuart, and Martin Libicki (eds.), *Dominant Battlespace Knowledge*, Washington D.C.: Institute for National Strategic Studies, National Defense University Press, 1995.

Jones, Carl, and Barry Wilson, unpublished research on what's new in *JICM 1.5,* Santa Monica, Calif.: RAND.

Keaney, Thomas, and Eliot Cohen, *Gulf War Air Power Survey: Summary Report,* Washington, D.C.: Department of the Air Force, 1993.

Khalilzad, Zalmay M., "The United States and the Persian Gulf: Preventing Regional Hegemony," *Survival*, 1995.

Khalilzad, Zalmay M. (ed.), *Strategic Appraisal 1996*, Santa Monica, Calif.: RAND, MR-543-AF, 1996.

Khalilzad, Zalmay M., and David A. Ochmanek, "Rethinking US Defence Planning," *Survival*, Vol. 39, No. 1, 1997a, pp. 43–64.

Khalilzad, Zalmay M., and David A. Ochmanek (eds.), *Strategic Appraisal 1997: Strategy and Defense Planning for the 21st Century*, Santa Monica, Calif.: RAND, MR-826-AF, 1997b.

Khalilzad, Zalmay M., David A. Shlapak, and Daniel Byman*, The Implications of the Possible End of the Arab-Israeli Conflict for Gulf Security*, Santa Monica, Calif.: RAND, MR-822-AF, 1997.

Knorr, Klaus, and Patrick Morgan, *Strategic Military Surprise: Incentives and Opportunities*, New York: National Strategy Information Center, 1983.

MacGreggor, Douglas A., *Breaking the Phalanx: A New Design for Landpower in the 21st Century*, Westport, Conn.: Praeger, 1997.

Matsumura, John, Randall Steeb, Tom J. Herbert, Mark R. Lees, Scot Eisenhard, and Angela B. Stich, *Analytic Support to the Defense Science Board: Tactics and Technology for 21st Century Military Superiority*, Santa Monica, Calif.: RAND, DB-198-A, 1997.

Moore, Louis, unpublished RAND research on power projection in the 21st century.

National Research Council, *Technology for Future Naval Forces: The United States Navy and Marine Corps, Becoming a 21st Century Force,* 1997.

———, *The Navy and Marine Corps in Regional Conflict in the 21st Century*, Washington, D.C.: National Academy Press, 1996.

O'Hanlon, Michael, *Defense Planning for the Late 1990s: Beyond the Desert Storm Framework*, Washington, D.C.: Brookings, 1995.

Panel on Analysis and Modeling, *Understanding Distributed-Force Concepts for Rapid Deployment Operations to Thwart Aggression*, Washington, D.C.: Defense Science Board, 1996.

Perry, William, *Annual Report to the President and Congress*, Department of Defense, March 1996.

Report of the National Defense Panel, Department of Defense, 1997.

Shalikashvili, General John, *Joint Vision 2010*, Joint Chiefs of Staff, Department of Defense, 1996.

Steeb, Randall, John Matsumura, Terrell G. Covington, Thomas J. Herbert, and Scot Eisenhard, *Rapid Force Projection: Exploring New Technology Concepts for Light Airborne Forces*, Santa Monica, Calif.: RAND, DB-168-A/OSD, 1996.

Treverton, Gregory, and Bruce W. Bennett, *Integrating Counterproliferation into Defense Planning*, Santa Monica, Calif.: RAND, IP-158, 1997.